Spinoff 1996

National Aeronautics and Space Administration

Office of Space Access and Technology

Commercial Development and Technology Transfer Division

by James J. Haggerty

National Aeronautics and
Space Administration

Foreword

The year 1996 has been a particularly fruitful time for NASA. The Galileo spacecraft is conducting a comprehensive investigation of Jupiter from uniquely productive vantage points in orbit around the giant planet. The Hubble Space Telescope continues to astound the world with its frequent revelations of space phenomena. Space Shuttle/*Mir* dockings have been carried out with impressive professionalism by the American and Russian crews. Accelerating deliveries of flight hardware has kept the International Space Station on track for initial assembly operations next year. And the aeronautics program marked its "return to flight," a renewed emphasis on the use of experimental aircraft with the unveiling of the X-36 tailless experimental vehicle.

We, in NASA, are justifiably proud of these and other accomplishments in aeronautics and space. We are equally proud of another type of accomplishment, one which is less visible to the public but just as important to the future of aerospace in America as are our dramatic operational achievements: we are reinventing NASA to better structure its organization and activities to meet the realities of the times and its potential for the future.

The dictates of reducing the national deficit have made it necessary for the agency to absorb large-scale funding cuts. The challenge has been to cope with these reductions while maintaining program stability. We have responded by seeking and finding new efficiencies in every phase of NASA operations to achieve cost savings with minimal loss of capability. Necessity, in this instance, is the mother of reinvention.

We have reversed cost growth; whereas NASA was experiencing large cost overruns four years ago, we are holding the line on meeting our costs estimates, and in some cases, underrunning those estimates. We have generated considerable savings by streamlining the management of the International Space Station and by restructuring several other large space programs. We have created many processes to reduce costs within the Space Shuttle program without compromising our high safety standards. We are realizing markedly significant savings by compressing the time it takes to design and develop a spacecraft vehicle with costs as a major consideration.

These are just a few of literally hundreds of examples wherein NASA is effecting major savings while maintaining the integrity and capabilities of the programs.

Despite this sharp focus on economy, NASA has by no means abandoned the exciting goals envisioned throughout its heritage. We have developed a NASA Strategic Plan that looks well into the future and targets such possibilities as human planetary exploration missions, lunar-based observatories, hypersonic aircraft, space-based commerce, and a global system for monitoring Earth's land, seas, and atmosphere. And we have created roadmaps that define the technologies needed to attain those goals.

Our strategic plan is bold and exciting, yet a pragmatic plan that accepts the probability of stringent funding for some time to come. We know that, if we are to attain the goals, we will have to do more with less, do it faster, and do it better. We are confident we can do that — because we're doing it now.

Daniel S. Goldin
Administrator

National Aeronautics and
Space Administration

Introduction

By their challenging nature, NASA programs are particularly demanding of technological input. Meeting the aeronautical and space goals of the past four decades has necessitated leading edge advancements across a diverse spectrum that embraces virtually every scientific and technological discipline.

Technology is simply knowledge and, like other forms of knowledge, it is often broadly applied and transferable. For that reason, the vast storehouse of technology NASA has built is a national resource, a bank of knowledge available for commercial applications and enhancements to the quality of life—"spinoff"—to new products and processes of benefit to the national economy, industrial efficiency and human welfare.

Multiple use of technology has never been more important. Budgetary stringency is reducing the amount of government funding available for new research and development, but at the same time intensifying international competition demands increasing technological innovation to strengthen the U.S. posture in the global marketplace. Reuse of technology offers a relatively inexpensive supplementary means of partnering with industry focused on bringing new products and processes to the market.

More than a thousand of spinoff products and processes have emerged from reapplication of technology developed for NASA mission programs. Each has contributed some measure of benefit to the national economy, productivity or lifestyle; some bring only moderate increments of gain, but many generate benefits of significant order with economic values in the millions of dollars.

Other technologies with moderate economic return have added measurably to the quality of life of U.S. citizens. Collectively, they represent a substantial dividend on the national investment in aerospace research.

By Congressional mandate, it is NASA's responsibility to promote expansion of spinoff in the public interest. Through its Technology Transfer Program, NASA seeks to encourage greater use of its technological resources by providing a link between the technology and those who might be able to put it to advantageous use. The program's aim is to broaden and accelerate the transfer accomplishments and thereby to gain national benefit in terms of new products, services, and new jobs.

This publication is an instrument of—and documents the outcome of—that purpose. It is intended to heighten awareness of the technology available for transfer and its potential for public benefit.

Spinoff 1996 is organized in three sections:

Section 1 summarizes NASA's current mainline programs, whose objectives require development of new technology and therefore replenish and expand the bank of knowledge available for reapplication.

Section 2, the focal point of this volume, contains a representative sampling of spinoff products and processes that resulted from secondary application of NASA technology.

Section 3 describes the various mechanisms NASA employs to stimulate technology transfer and lists, in an appendix, contact sources for further information about the Technology Transfer Program.

I hope you enjoy reading about NASA's newest spinoffs.

Robert L. Norwood

Dr. Robert L. Norwood
*Director, Commercial Development
and Technology Transfer Division*
National Aeronautics and
Space Administration

Contents

Aerospace Aims

Technology Twice Used

Technology Transfer

Aerospace Aims

An illustrated summary of NASA's major aeronautical and space programs, their goals and directions, their contributions to American scientific and technological growth, and their potential for practical benefit

Space Operations

U.S./Russian cooperative missions highlight development of the International Space Station

On March 24, 1996, the Space Shuttle Orbiter *Atlantis* flew to a rendezvous with the Russian space station *Mir*, docked with it, and the six-astronaut crew of Shuttle flight STS-76 joined the three cosmonaut crew of the orbiting station.

Among the Americans was mission specialist Shannon W. Lucid, a veteran of five Shuttle flights who was reporting for duty aboard *Mir* as a cosmonaut researcher. Lucid was to spend almost six months on *Mir*, then return to Earth on another Shuttle/*Mir* rendezvous/docking mission. Her arrival at *Mir* kicked off a two-year span of continuous U.S. presence in orbit.

STS-76 marked the third docking flight with *Mir* and the fifth in a multistep preliminary agenda intended to provide a framework for

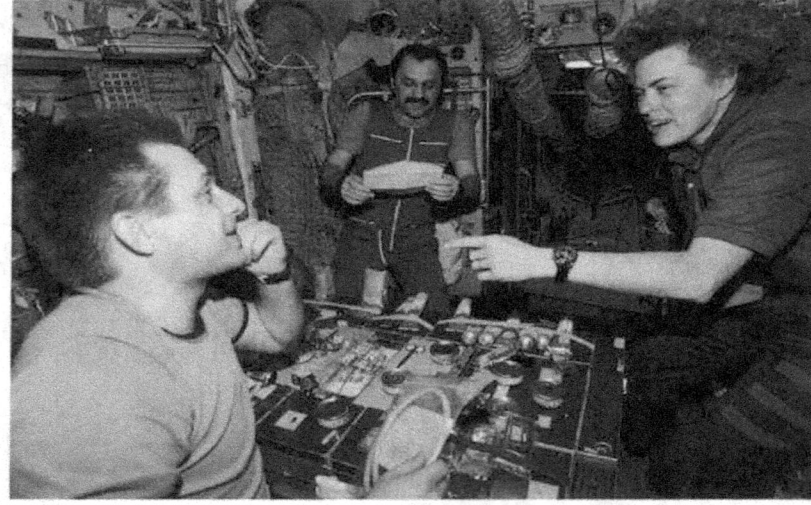

U.S. astronaut Shannon Lucid helps her cosmonaut crewmates inventory newly-delivered food supplies in the base block module of Russia's **Mir** *space station. At left is* **Mir** *commander Yuri Onufriyenko and in the background is flight engineer Yuri Usachev.*

International Space Station assembly operations, which begin in 1997. Known as Phase I of the International Space Station developmental program, this agenda is intended to lay the foundation for Phases II/III through integration of U.S. and Russian hardware, working methods and scientific procedures; risk reduction, or mitigation of potential surprises; and early initiation of joint scientific and technological research. Prior flights included:

• STS-60, Orbiter *Discovery*, launched February 11, 1994, in which cosmonaut Sergei Krikalev served on the Shuttle crew as a mission specialist.

• STS-63, Orbiter *Discovery*, launched February 3, 1995 to a rendezvous with *Mir* and a slow "flyaround," in which the Orbiter circled the station at a distance of about 450 feet for photography and communications tests. The rendezvous and flyaround validated a number of techniques for subsequent employment in docking missions, such as those essential to orbital assembly of the International Space Station.

• STS-74, Orbiter *Atlantis*, launched November 12, 1995 to the second Shuttle/*Mir* docking. The flight focused on delivery of equipment to *Mir* and demonstration of the Russian-built docking system. STS-74 underlined the international flavor of the space station effort: the Shuttle crew was made up of four Americans and a Canadian mission specialist; hardware in the payload bay included the U.S.-built Orbiter Docking System, the Russian docking module, the Canadian Remote Manipulator System, and two Russian-built solar arrays. The *Mir* was staffed by two Russian and one German cosmonauts, who were operating Russian and European Space Agency equipment.

The crew of STS-74 (two-tone shirts) join the crew of the Mir space station for a group portrait after their November 1995 Shuttle/Mir linkup. The Shuttle members are, from right to left, STS-74 commander Kenneth D. Cameron; pilot James D. Halsell Jr.; Canadian mission specialist Chris Hadfield; NASA mission specialists Jerry L. Ross and William S. McArthur Jr.

On STS-76, *Atlantis* and *Mir* remained docked for five days during which the crews transferred to *Mir* some 1,500 pounds of water and two tons of scientific equipment, and moved to *Atlantis* a number of experiment samples and miscellaneous equipment. They also conducted joint experiments with the ESA's Biorack experiment rack housed in a SPACEHAB pressurized module aboard the Orbiter; the Biorack work embraced 11 separate investigations of the effects of microgravity on a variety of living materials. Another highlight was the first U.S. EVA (extravehicular activity) around two mated spacecraft, a six-hour "walk" performed by mission specialists Linda M. Goodwin and Rich Clifford. STS-76 landed at Edwards Air Force Base, California on March 31, 1995.

At *Spinoff* publication time (midyear 1996), the Orbiter *Atlantis* was poised for its fourth docking with *Mir* in mid-August. It was to pick up Shannon Lucid and deliver her replacement, astronaut John Blaha. After that, the Phase I plan contemplated three more Shuttle/*Mir* hookups in December 1996, May 1997 and September 1997. Each flight will involve pickup of the American crew member of *Mir* and delivery of a replacement; all flights will be made by *Atlantis*, the only Orbiter equipped for docking with *Mir*.

(Continued)

International Space Station

The International Space Station will be a permanent laboratory for human-monitored long term research in the unique environment of Earth-orbital space, an environment that cannot be duplicated on Earth for long duration experiments. Research at the station will focus on two key areas: life sciences and materials sciences.

Life science research is expected to lead to a clearer understanding of basic processes to provide a foundation for development of advanced medications for improved human health care. Material research offers promise of improved metals, composites and plastics for significant advances in technologies for communications, transportation and a broad range of industrial processing operations.

The International Space Station draws upon the resources and scientific/technological expertise of 13 cooperating nations, including the U.S., Canada, Japan, Russia and nine nations of the European Space Agency (Belgium, Denmark, France, Germany, Italy, Norway, Spain, The Netherlands and the United Kingdom). The prime contractor is The Boeing Company and the principal subcontractors are McDonnell Douglas Corporation and the Rocketdyne Division of Rockwell International.

While the Phase I flight program was under way in 1995/96, manufacturers were turning out the first hardware components of the space station. Among major segments completed in 1995 were two Boeing-built nodes (Node 1 and Node 2). The nodes will serve as connecting passageways between modules. Node 1 will be the first U.S.-built hardware delivered to orbit.

Also built by Boeing is the structure for the U.S. laboratory module where astronauts will perform continuous scientific research; it was delivered early in 1996.

Completed early in 1996 was the Boeing-built structure for the U.S. Laboratory Module, a key element of the International Space Station. The aluminum module is 28 feet long and weighs three tons.

Phase II of the space station program, construction in orbit, begins in November 1997 with the launch on a Russian *Proton* vehicle of the FGB functional cargo block. The FGB is a 21-ton element, built in Russia but purchased by the U.S., that will provide attitude control and propulsion during the early assembly operations, plus solar power and berthing ports for additional modules.

A month later, Node 1 will be delivered by the Space Shuttle and attached to the FGB. In May 1998, the embryo space station will grow with the addition of the *Proton*-boosted Russian Service Module, which provides life support and habitation facilities, utilities and thrusters. Shortly thereafter, in May 1998, the crew

transfer vehicle—a Russian Soyuz TM capsule—will be joined to the station. In June 1998, the first three-person crew will begin its orbital stay.

Further additions to the expanding station in the latter part of 1998 and early 1999 will include one of the four U.S. solar array modules, which will provide about 23 kilowatts of power; segments of the central truss; the U.S. Laboratory Module; the Canadian-built mobile servicing system; the Russian Universal Docking Module; and the equipment for outfitting the U.S. Laboratory Module. With Shuttle delivery and attachment of an airlock in the spring of 1999, Phase II officially comes to a close.

(Continued)

The interim International Space Station will look like this. In the right foreground is the U.S. Laboratory Module and the station's airlock. In the center of the horizontal string of modules is the FGB energy block. The solar power array at top is one of four that will provide power for the complete station. Below the tower is the Russian-built Universal Docking Module and, at bottom, one of two crew transfer vehicles.

International Space Station *(Continued)*

In Phase III, the International Space Station will progress gradually to its ultimate status as a fully operational permanent orbital research facility. Among key additions to the core configuration are the remaining modules of the U.S.-built solar array; the Japanese Experiment Module (JEM), to be delivered in 2000; and the U.S. habitation module (February 2002), which contains the galley, toilet, shower, sleep stations and medical facilities.

With the delivery of a second Russian crew transfer vehicle in June 2002, the station will be virtually complete (the European Space Agency (ESA) laboratory, known as the *Columbus* Orbital Facility, will be joined to the station early in 2003). At that point, the station will have a full six-person crew capability.

The completed station will measure 361 feet from tip to tip of the solar arrays. That corre-sponds to the length of a football field with both end zones included. However, the area of the station complex is equal to that of two football fields.

The pressurized living and working space is roughly equivalent to the passenger cabin volume of two Boeing 747 jetliners. The atmospheric pressure within the pressurized modules will be 14.7 pounds per square inch, same as on Earth's surface.

There will be seven laboratories. The U.S. is providing two of them, the basic laboratory module and a special Centrifuge Accommodation Module. There will be three Russian research modules, the Japanese JEM and ESA's *Columbus* module. The U.S., ESA and Japanese laboratories together provide 33 International Standard Payload Racks; additional payload capability will be available in the Russian

A view of the International Space Station in its final configuration with a Space Shuttle Orbiter docked at the fore port. The cylinder near the Orbiter's nose is the U.S. Centrifuge Accommodation Module. Below it, hidden by the Orbiter, is the U.S. lab module, flanked by the European (left) and Japanese laboratories.

modules. In addition, the JEM has an exposed "back porch" with 10 mounting spaces for experiments that require long duration contact with the space environment; the JEM has a small robotic arm for moving back porch payloads.

A central girder connecting the various modules and the main solar power array is the U.S.-built integrated truss. Moving along the truss for robotic assembly and maintenance operations is the Canadian-built Remote Manipulator System with its 55-foot robot arm and mobile transporter. The four modules of the solar array, generating a combined 92 kilowatts of power, rotate on the truss to maximize their exposure to the Sun.

The International Space Station will operate at an average altitude of 220 miles. At that altitude, minute drag forces will cause the station to lose height very gradually, so it will be necessary to reboost it every 90 days. The reboosting will be accomplished by the FGB.

Beginning with the 1997 launches of the Russian *Proton* and the Space Shuttle, there will be 73 assembly and service flights until the station becomes fully operational in midyear 2002. The Space Shuttle will make 27 trips, 21 for assembly operations and six for utilization/outfitting. The Russian *Proton* and *Progress* and Ukrainian *Zenit* launch vehicles will make 45 flights; 15 of them will be made by the *Progress* vehicle, bringing up propellant for the reboost work.

This concept views the station from the opposite (aft port) end. In the foreground (lower right) is the Russian Service Module, with living and working room for three crew members. Next, toward center of photo, is the FGB energy block, then (near the Orbiter) the U.S. lab module. The vertically-mounted cylinder below it is the U.S. habitation module.

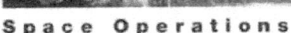

Space Shuttle Operations

In addition to the Shuttle/*Mir* docking flights (see pages 2-3), NASA conducted several important Space Shuttle missions in 1995/96 whose main objectives were scientific/technological rather than preliminaries to International Space Station assembly.

After four highly successful flights earlier in 1995, NASA launched STS-69, Orbiter *Endeavour*, on September 7 for a very demanding mission in which—for the first time—two separate payloads were deployed from the Orbiter and later retrieved.

The first was Spartan 201, the third Spartan mission in a planned series of four. Deployed

STS-69 astronauts James S. Voss (foreground) and Michael L. Gernhardt evaluated tools and techniques designed for on-orbit assembly of the International Space Station.

on mission day two, Spartan 201 carried two scientific instruments for study of the Sun's outer atmosphere (corona) and its transition into the solar wind that constantly flows past Earth. Both instruments operated smoothly and both returned good data.

On flight day five, the *Endeavour* crew deployed the Wake Shield Facility-2 (WSF-2), a 12-foot-diameter disc designed to create an "ultravacuum" environment for growing electronics materials of significantly higher quality than can be produced on Earth. During its period of free flight, WSF-2 was able to perform four of seven planned thin film growth runs, despite attitude control problems.

On flight day 10, astronauts James S. Voss and Michael L. Gernhardt conducted a six-hour-46-minute spacewalk. They tested a variety of tools and techniques that may be employed in assembly of the International Space Station and evaluated improvements made to their extravehicular activity (EVA) suits. The STS-69 crew also covered a lengthy agenda of other experiments, including growth of optical crystals and thin films with commercial potential plus a variety of astronomical observations.

Shuttle mission STS-73, Orbiter *Columbia*, launched October 20, 1995, marked the second flight of the U.S. Microgravity Laboratory (USML), first flown in 1992. USML-2 built on the foundation of the prior mission, focusing in the same general areas of research, with many experiments flying for the second time. The Orbiter crew divided into two teams to work around the clock in the 23-foot-long pressurized Spacelab module mounted in the Orbiter's payload bay.

Research was conducted in five areas: fluid physics, materials science, biotechnology, combustion science and commercial space processing. Flying for the first time was a combustion experiment in which more than 25

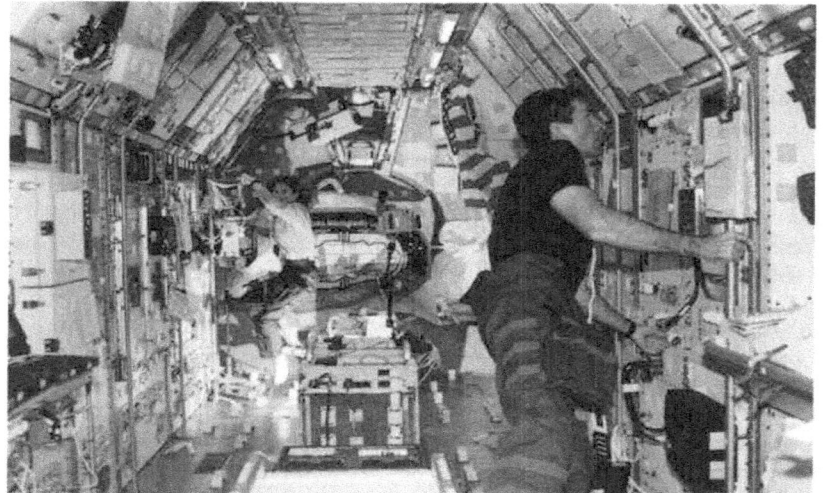

This wide view provides a perspective of the working environment in the U.S. Microgravity Laboratory, flown aboard STS-73. At left, working at the "glovebox," is mission specialist Catherine G. Coleman, and at right is payload specialist Fred Leslie, who is conducting a fluid physics experiment.

droplets of a variety of fuels were ignited for study of how fuels burn in microgravity. In the Astroculture Plant Growth Facility, which is set to become a commercial system, small potatoes were grown from tubers, demonstrating that edible foods can be grown in space. A number of other experiments involved growth of crystals for post-mission study; crystals grown in orbit are generally larger and of higher quality than crystals grown in Earth facilities, offering potential for use in development of advanced pharmaceuticals or computer chips that are faster and use less power than traditional chips. STS-73 landed at Kennedy Space Center (KSC) after almost 16 days in orbit.

The year's final mission, STS-74, Orbiter *Atlantis*, launched November 12, was devoted entirely to docking with the Russian space station *Mir* and subsequent joint U.S./Russian experimentation (see pages 2-3). STS-74 landed at KSC on November 20, rounding out a highly successful 1995 Shuttle agenda of seven missions.

(Continued)

Working on experiments aboard STS-73 are payload commander Kathryn C. Thornton (foreground) and payload specialist Albert Sacco Jr.

Space Shuttle Operations *(Continued)*

The first Space Shuttle flight of 1996 was STS-72, Orbiter *Endeavour*, launched January 11, a nine-day mission highlighted by retrieval of a Japanese satellite that had spent 10 months in orbit, deployment and retrieval of a NASA science payload, and two spacewalks.

On flight day three, mission specialist Koichi Wakata of the National Space Development Agency of Japan operated the Shuttle's remote manipulator arm to pluck the Japanese Space Flyer Unit from orbit. On the following day, Wakata again used the robot arm, this time to deploy the NASA OAST-Flyer (Office of Aeronautics and Space Technology), a free-flying platform with four experiments: one testing the accuracy of computer models for predicting spacecraft exposure to contamination; one demonstrating satellite-aided global positioning; one testing laser ordnance devices; and an amateur radio communications experiment. The OAST-Flyer was successfully retrieved on flight day six.

Two extravehicular activities (EVAs) were conducted by mission specialists Winston Scott, Daniel T. Barry and Leroy Chiao as part of a continuing series of preliminaries to on-orbit assembly of the International Space Station. They evaluated a new portable work platform and a space station utility box designed to hold avionics and fluid line connects. Scott also tested the warmth of the EVA suit in temperatures as low as minus 104 degrees Fahrenheit.

Other experiments included use of a cargo bay-mounted instrument for measuring ozone concentrations in the atmosphere, experiments related to protein crystal growth, a National Institutes of Health effort to validate models of microgravity effects on human bones, muscles and cells, and a separate study of the effects of microgravity on rodents.

On February 22, NASA launched STS-75, Orbiter *Columbia*, on a reflight of the U.S./Italian Tethered Satellite Systems (TSS). The TSS is a tool for

This "fisheye" view shows astronaut Winston Scott conducting a spacewalk during STS-72 in January 1996. The stowed spacecraft in the rear of the Orbiter's payload bay are the Japanese Space Flyer Unit, retrieved after 10 months in orbit, and NASA's OAST-Flyer, retrieved after a two-day free flight.

studying the power generation potential of a tethered satellite operating in the electrically-charged ionosphere. Deployed on flight day three, the TSS was gathering excellent data when the pencil-thin tether snapped; the satellite had almost reached full deployment 12.5 miles from the Orbiter.

The other primary payload on STS-75 was the U.S. Microgravity Payload-3 (USMP-3), which embraced a wide range of U.S. and international experiments in crystal growth and materials processing. USMP-3 performed nominally.

The year's third mission was STS-76, Orbiter *Atlantis*, launched March 22. The flight was devoted primarily to rendezvous and docking with the *Mir* space station, but it also included a lengthy agenda of U.S./Russian scientific experiments (see pages 2-3).

On May 19 NASA launched the fourth flight of 1996, STS-77 Orbiter *Endeavour*. Primary payloads included commercial space product development payloads flying aboard the pressurized, commercially-developed SPACEHAB module (see page 24); an inflatable antenna experiment carried by the Spartan free-flying spacecraft; and four experiments called TEAMS (Technology Experiments for Advancing Missions in Space). TEAMS included an effort to determine the degree of accuracy with which the Global Positioning System can supply attitude information to a spacecraft; a test of improved methods for in-space refueling; an evaluation of liquid metal heat pipes in microgravity; and a demonstration of aerodynamic stability in the upper atmosphere.

At *Spinoff* publication time, STS-79, Orbiter *Atlantis* was being readied for mid-August launch to a fourth docking with the Russian *Mir*. A November mission, STS-80, Orbiter *Columbia*, was to feature the third flight of the WSF and in December *Atlantis* was to dock with *Mir* for the fifth time on STS-81.

This traditional in-flight crew portrait emphasizes the international character of Shuttle flight STS-75. At bottom center is mission commander Andrew M. Allen. Clockwise from him are payload commander Franklin R. Chang-Diaz, NASA; Maurizio Cheli and Claude Nicollier, representing the European Space Agency; pilot Scott Horowitz, NASA; payload specialist Umberto Guidoni, Italian Space Agency; and Jeffrey A. Hoffman, NASA mission specialist.

A view of the Tethered Satellite at the start of its deployment from STS-75. The satellite was lost when the tether snapped but the experiment provided considerable useful data.

Probing the Universe

NASA's space

science program

seeks knowledge

about Earth

and its place

in the universe

Asteroids are small celestial bodies ranging in size from a mile or two in diameter to almost 500 miles. They orbit the Sun in the same direction as the principal planets, although their orbits, as a rule, are far more eccentric; most revolve about the Sun in a cosmic zone between the orbits of Mars and Jupiter.

Asteroids have on occasions in the distant past collided with Earth and left evidence of great devastation; one theory holds that an asteroid impact with Earth caused the demise of the dinosaurs. These mysterious cosmic objects clearly have potential for influencing the evolution of the atmosphere and life on Earth.

This February 1996 launch from Kennedy Space Center sent the Near Earth Asteroid Rendezvous (NEAR) spacecraft off on a three-year flight to a rendezvous with a 25-mile-long asteroid known as 433 Eros.

They also offer clues to the nature of the early solar system processes and conditions, which have long been erased by evolution on Earth and other large bodies, but which are preserved in various forms on these small bodies that do not have sufficient gravitational power to retain an atmosphere.

All of which makes asteroids prime targets for comprehensive scientific investigation. Such an investigation is under way; before the century ends, scientists will know a great deal more about asteroids than they do today.

The investigation formally began on February 17, 1996 with the launch of the Near Earth Asteroid Rendezvous (NEAR) spacecraft on a three-year odyssey toward a rendezvous with a 25-mile-long asteroid known as 433 Eros. In February 1999, a firing of NEAR's main thruster will put the spacecraft into an orbit around the asteroid—the first spacecraft ever to accomplish such a feat—to circle Eros for a year, coming at times as close as 10 miles from its surface.

An international project conducted in cooperation with Germany and France, NEAR is designed to image the asteroid and measure its size, shape, volume, mass, gravity field and spin; its elemental and mineral composition; its mass distribution and magnetic field. Built by Johns Hopkins University's Applied Physics Laboratory (APL), the spacecraft measures approximately nine feet in length, five feet in diameter and weighs some 1,800 pounds. APL is NASA's manager for the project.

NEAR is an exciting and important project in its own right, but it takes on added significance by virtue of its status as the prototype, or pathfinder, of a relatively new NASA program known as Discovery, an ongoing effort to foster development of low cost spacecraft that will enable frequent solar system exploration missions. As the first flight project of that program, NEAR bears the responsibility for demonstrating that the concept of low-cost planetary exploration is valid.

NEAR was built under a ceiling of $150 million, a fraction of the cost of some interplanetary missions, and it was completed in only two years under a new streamlined management approach. Although cost was a primary consideration throughout the development period, NASA and APL did not compromise capability in the search for economies. NEAR has a full complement of five sophisticated instruments and multiple redundancies in its operating systems.

NEAR exemplifies one aspect—solar system exploration—of NASA's broad space science program, whose primary missions and goals embrace these general areas of investigation:
• Examining the content, structure, origin and evolution of the galaxy and the universe;
• Defining the relationships among the Sun, Earth and the heliosphere;
• Seeking greater understanding of the origin and evolution of planetary systems;
• Pursuing greater understanding of the origin and distribution of life in the universe;
• Observing Earth's air, land, water and life resources toward greater understanding of the complex mechanisms that control Earth's behavior.

Selected examples of ongoing and upcoming programs in these areas are contained in the following pages.

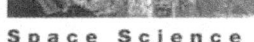

Solar System Exploration

The planetary systems research segment of NASA's space science program centers largely on exploration of our solar system but NASA's latest strategic plan includes "a comprehensive search for planets and planetary formation around other stars." Among the other near term goals are to complete reconnaissance of the solar system; conduct orbital surveys and begin surface exploration of the most fascinating and most accessible bodies; and to complete the inventory of all near-Earth objects measuring one kilometer or more.

The long term goals are to go beyond "reconnaissance" of our solar system to detailed surveys, including sample analysis, of the most important bodies, and to identify planets around other stars that might be habitable.

The major planetary systems program in flight status is Galileo, a two-part vehicle consisting of a spacecraft and an instrumented probe designed for a comprehensive examination of Jupiter. Launched in 1990, Galileo arrived at the giant planet in December 1995 and swung into an interim orbit around Jupiter. Among Galileo's early discoveries, its instruments uncovered evidence that Jupiter's moon Io, the most geologically active body in the solar system, has its own magnetic field; if so, it would be the first planetary moon known to have one.

Meanwhile, Galileo's atmospheric probe plunged into the Jovian atmosphere on December 7, 1995 to investigate the planet's composition and physical state. Carrying six instruments, the probe descended by parachute for 57 minutes, reporting a series of "unexpected and often startling" discoveries and relaying its data to the orbiting main spacecraft. The data is undergoing extensive study.

On March 14, 1996, Galileo's main engine was fired to boost the spacecraft to a higher orbit and move it further away from the harmful

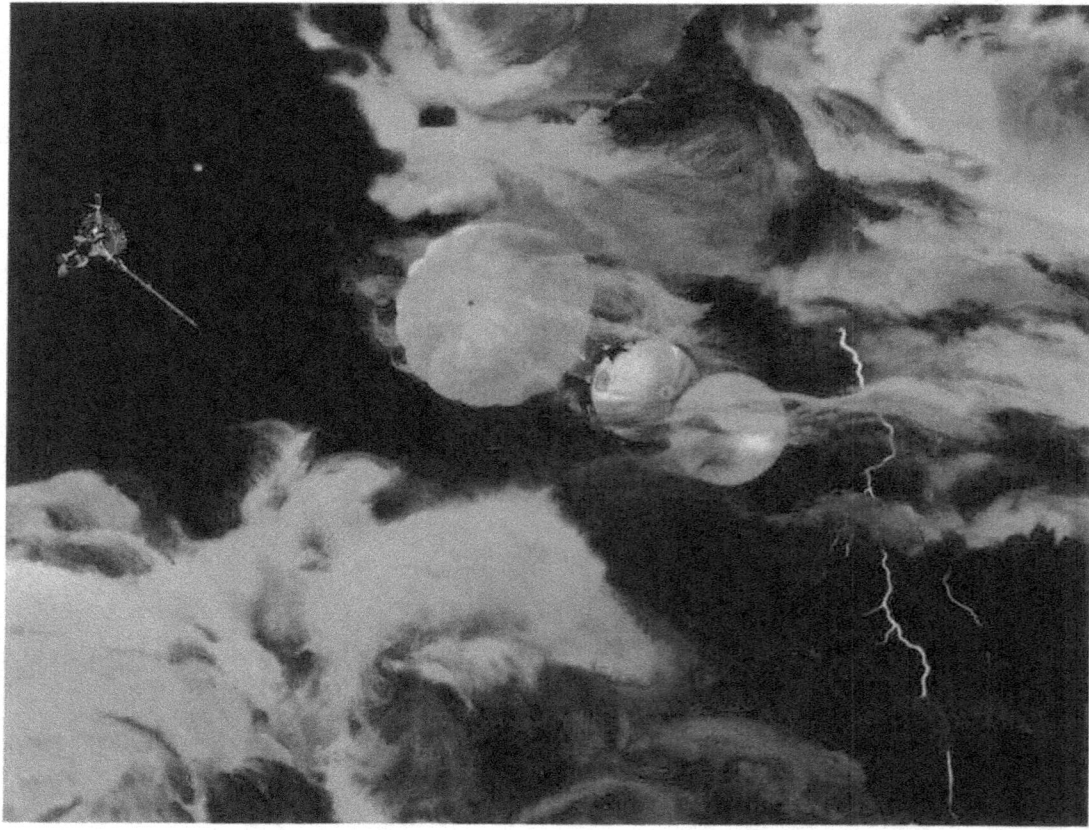

An artist's conception of the Galileo probe (center) descending through the atmosphere of Jupiter as the Galileo main spacecraft orbits the planet (upper left). The probe relayed a wealth of data for 57 minutes before it was destroyed by Jupiter's immense gravity.

radiations emanating from the giant planet. That put the spacecraft into a final orbit that will enable it to study in detail the planet's four largest moons—Io, Ganymede, Europa and Callisto—and record extensive data about Jupiter itself during 11 orbits over a 23-month span. Galileo is a cooperative U.S./Germany project managed for NASA by Jet Propulsion Laboratory (JPL). The main spacecraft was designed and built by JPL; Ames Research Center has management responsibility for the probe, which was built by Hughes Aircraft.

At *Spinoff* publication time, NASA was preparing to resume the comprehensive exploration of Mars that began with the Viking orbiters/landers of the late 1970s. Scheduled for launch in November 1996, the Mars Global Surveyor carries five instruments to conduct a systematic mapping of Mars and to obtain extensive data on the geophysical/climatological history of the planet and the evolution of its interior and surface. After reaching Mars in 1997, the spacecraft will conduct its primary mission from polar orbit for two years, then serve as an orbiting communications station for another three years, relaying data from follow-on U.S. and international Mars landers. JPL is program manager; Lockheed Martin Astronautics built the spacecraft.

Under NASA's Discovery program, an effort to develop frequent, small planetary missions that perform high quality scientific investigations at low cost while emphasizing involvement by the academic and research communities, NASA plans to send a Mars lander to the planet in July 1997. Managed by JPL and called Mars Pathfinder, the spacecraft contains a surface rover and three science instruments for acquiring geological/meteorological data and conducting technology experiments to pave the way for future low-cost robotic exploration of Mars.

Other planned Discovery missions include the Lunar Prospector (mid-1997 launch), which will map the Moon's surface composition, magnetic fields and gravity fields from low-altitude orbit; and Stardust (1999), a mission involving rendezvous with a comet and return of an interstellar dust sample. Both spacecraft are built by Lockheed Martin.

The principal planetary mission being readied is Cassini, a joint project of NASA, the European Space Agency (ESA) and the Italian Space Agency, managed for NASA by JPL. The flight vehicle consists of the main Cassini spacecraft and the ESA-built Huygens Probe, a 750-pound, six-instrument package that will descend into the atmosphere of Saturn's moon Titan, which is believed to be chemically similar to the atmosphere of early Earth and is therefore of immense scientific interest.

To be launched in October 1997, Cassini will make flybys of Venus and Jupiter en route to a rendezvous with Saturn in July 2004, where it will be inserted into a loose elliptical orbit. Cassini will release the Huygens Probe during the first orbit, then make approximately 40 revolutions over a span of four years, while the spacecraft's 12 instruments conduct a detailed exploration of the whole Saturnian system, including Titan and the planet's other icy moons.

This artist's concept shows the Cassini spacecraft orbiting around Saturn, just after deploying a probe that will descend into the atmosphere of Saturn's moon Titan. To be launched in 1997, Cassini will reach Saturn in July 2004 and orbit the planet for four years thereafter.

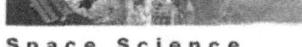
The Galaxy and the Universe

What is the universe? How did it come into being? How does it work? What is its ultimate fate? These are some of the fundamental questions addressed by the galaxy/universe research segment of NASA's space science program. The long term goal is to provide the answers to those questions within 25 to 50 years. The near term goals are to complete development of an initial observational capability; complete the survey of cosmic rays and interstellar gas (examples of extrasolar matter); and carry out basic new tests of gravitational theory.

Astronomy has made giant strides since the advent of satellite-based telescopes in the 1960s. Operating above the obscuring atmosphere, these instruments provide undistorted views of the universe, and they can observe in bands of the electromagnetic spectrum— ultraviolet, infrared, x-ray and gamma ray, for example—to pick up star radiations normally absorbed or blocked out by the atmosphere and thus not detectable by ground telescopes. This latter capability is particularly important to astronomical science, because each band of the spectrum offers a different set of clues to the origin and evolution of the universe.

Since the 1960s, NASA has orbited a series of orbiting observatories of ever-increasing capability, culminating with the service debuts of the Hubble Space Telescope (1990) and the Compton Gamma Ray Observatory (1991). These two "Great Observatories" have been regularly providing astronomers major new discoveries about the cosmos.

For example, in 1995-96, the Hubble Space Telescope (HST) confirmed the presence of a second black hole in the universe; recorded the emergence of infant stars from dense, compact pockets of interstellar gas; and made the first unambiguous detection and imaging of a "brown dwarf," an object too massive and too hot to be a planet but too small and too

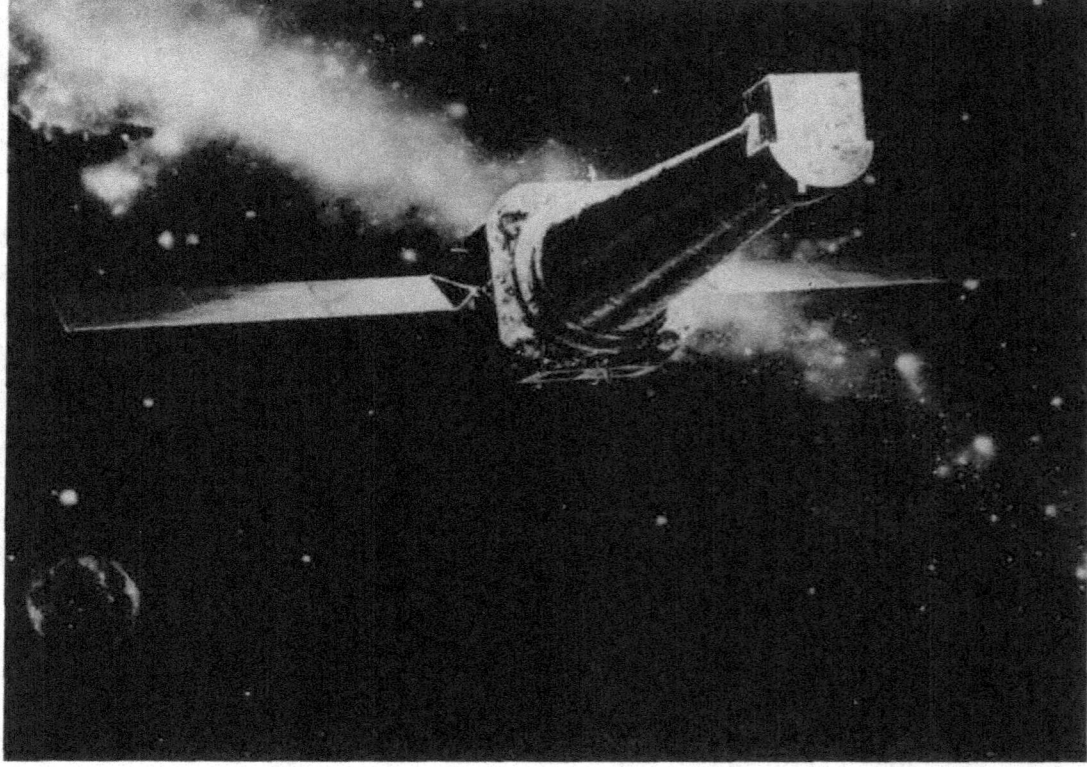

The Advanced X-ray Astrophysics Facility, third in NASA's Great Observatories series, will make its service debut in 1998.

cool to shine like a star. The brown dwarf is the faintest object ever seen orbiting a star.

The Compton Gamma Ray Observatory (CGRO), which had recorded more than 1,400 mysterious gamma ray bursts by the end of 1995, completed a survey of the highest energy gamma ray sources and showed that about half of them were quasars with beams of energy pointed directly at Earth; the remaining sources were not identifiable.

Designed for a 15-year lifetime, made possible by Space Shuttle servicing/reequipment missions, HST will continue to operate well into the 21st century. Since the telescope became operational, Goddard Space Flight Center has had responsibility for controlling the HST and processing its imagery and data; the data is collected and distributed by the Space Telescope Science Institute in Baltimore, Maryland.

Unlike Hubble, the CGRO is not designed for Shuttle servicing, but its lifetime can be extended to 10 years through occasional altitude reboosts by its on-board propulsion system. Goddard Space Flight Center is NASA's CGRO manager; TRW Inc. is principal contractor. International participation includes Germany, The Netherlands, United Kingdom and the European Space Agency (ESA). ESA and NASA are teaming in development of a CGRO successor, the International Gamma Ray Astrophysics Laboratory (INTEGRAL), planned for launch in 2001.

The third of the Great Observatories is the Advanced X-ray Astrophysics Facility, in development and scheduled for launch in September 1998. AXAF will address some fundamental science questions by obtaining x-ray images of such objects as neutron stars, black hole candidates, quasars and active galaxies. Project manager is Marshall Space Flight Center and TRW Inc. is principal contractor.

The fourth member of the Great Observatories family is the Space Infrared Telescope Facility, which is intended to conduct advanced investigations of prime interest targets developed by earlier infrared observatories; managed by Jet Propulsion Laboratory, it is in study status in anticipation of hardware development in 1998 and orbital service in 2002.

Another infrared observatory development is the Stratospheric Observatory For Infrared Astronomy (SOFIA), a joint project of NASA and the German Space Agency DARA. SOFIA is an airborne rather than an orbital observatory. It consists of a 2.5 meter telescope mounted in a specially modified Boeing 747SP transport; it is planned for initial operation in 2000. SOFIA will replace NASA's aging Kuiper Airborne Observatory, a telescope-equipped C-141 transport that has been in service since 1974. SOFIA's telescope will be three times the diameter and about 10 times more sensitive than the Kuiper system.

One other major mission in the galaxy/universe research segment is Gravity Probe B, being built by Lockheed Martin under the management of Marshall Space Flight Center.

In development for initial service in 2000 is the Stratospheric Observatory For Infrared Astronomy, a modified Boeing 747 carrying a 2.5 meter telescope.

Intended for launch in December 1999, Gravity Probe B will perform tests of two fundamental predictions of Albert Einstein's general theory of relativity. The spacecraft will carry four precisely-manufactured, golf-ball-size crystal spheres inside a large dewar (thermos container). The levitated spheres, isolated from heat influence, will be spun at a precise rate and trained on a reference star. The reference axis will be compared with the gyro spin axes with high precision; if Einstein's predictions relative to the warping of time and space are correct, each sphere's axis should drift slightly from that of the reference star. The relatively large spacecraft—3 tons—will operate in a 400-mile-high polar orbit for a design lifetime of 1.6 years.

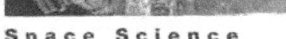

The Sun-Earth Connection

The Sun is a variable star, meaning that its activity varies over time. The changing degree of activity is due to two factors: its rotation and its convection processes, which transport hot gas from the solar interior to the surface. The interaction of these two motions—rotation and convection—generates powerful magnetic fields and influences the cyclic activity level demonstrated by the ebb and flow of sunspots and solar flares.

Additionally, activity in the Sun's corona—the white "halo" of gas seen during total eclipses—causes ejection at very high velocities of a hot electrified gas called the solar wind, which courses throughout the solar system transporting energy to Earth and all the other planets. Its interaction with Earth's magnetic fields causes a whole range of effects, such as the aurora magnetic storms, disruption of radio communications, and power surges in transmission lines.

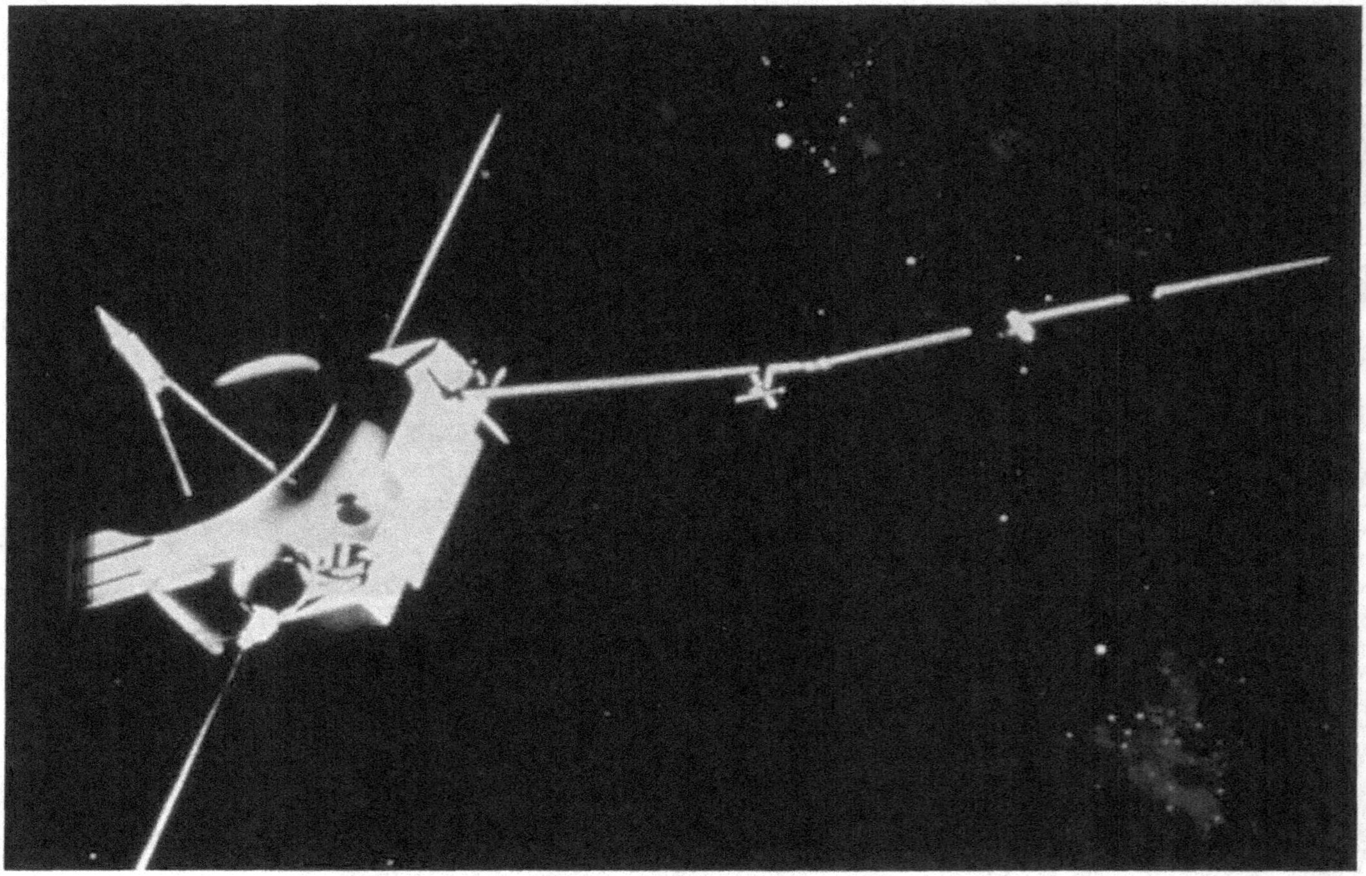

Ulysses became the first spacecraft to explore the Sun from polar orbit.

An important facet of NASA's space science program is solar-terrestrial research, which embraces the study of the Sun as a variable star, the origin and transmission of the solar wind, its interactions with Earth's magnetosphere, and how all these phenomena connect the Sun to the Earth and the heliosphere, the vaguely-bounded region of space where the Sun's magnetic field and the solar wind extend.

Over the years, NASA has employed a number of spacecraft to study the processes that link Earth with the Sun. In the 1990s, this activity has expanded under the International Solar Terrestrial Physics (ISTP) program, a joint effort of NASA, the European Space Agency (ESA) and the Japanese Institute of Space and Astronautics.

Not a component of ISTP but a joint NASA/ESA mission, Ulysses—launched October 6, 1990—is the first effort to explore the heliosphere from solar polar orbit over a full range of solar latitudes. After a roundabout four-year flight from Earth, Ulysses reached an area of the Sun's south pole in June 1994, then flew into the northern hemisphere of the heliosphere in 1995 and conducted a four-month observation of the north polar region. Late in 1995, Ulysses completed its primary mission, having returned volumes of invaluable data on the Sun's corona, wind, solar and non-solar cosmic rays, solar radio bursts and plasma waves, and the heliosphere's magnetic field. Ulysses is managed for NASA by Jet Propulsion Laboratory.

The Solar Heliospheric Observatory (SOHO), launched December 2, 1995, surprised scientists by its initial reports, which cited intense activity on the Sun's surface at a time—the low ebb of the 11-year solar cycle—when it should have been relatively inactive. A part of the ISTP program, SOHO is a joint NASA/ESA mission designed to perform remote measurements of the Sun and *in situ* measurements of the solar wind to improve knowledge of the corona and the origin of the solar wind.

The two-ton spacecraft was launched into orbit at a point about a million miles sunward from Earth, a point where the gravities of the Sun and Earth cancel each other out and provide a stable position from which to conduct long term continuous observation of the Sun. In the early months of 1996, SOHO returned to Earth motion pictures of the Sun's activity and a mass of data on solar interior dynamics and the composition of the solar wind. In addition to NASA and ESA, participants include Germany, the United Kingdom, France, Finland and Switzerland; Goddard Space Flight Center manages the NASA-provided elements of SOHO.

Another component of the ISTP program, launched into polar orbit on February 24, 1996, is the Polar spacecraft, built by Lockheed Martin Astro Space under the management of Goddard Space Flight Center. Polar and its sister spacecraft Wind (launched November 1, 1994) are a pair of complementary spacecraft, developed under NASA's Global Geoscience Space Program to gain broader understanding of the relationship between solar plasma emitted by the Sun and its interaction with Earth's magnetosphere, ionosphere and magnetic poles.

Polar's job is to measure the energy, energization and transport of plasma into the magnetosphere by the solar wind. Additionally, it is making direct measurements of global energy deposition into Earth's atmosphere. At midyear 1996, NASA was receiving good data from all 11 of Polar's instruments.

Two veteran spacecraft launched almost 30 years ago—Voyagers 1 and 2—are now playing a part in solar-terrestrial research. Having completed their grand tours of the solar system and flybys of the outer planets, they have been assigned new jobs: to observe cosmic rays and the solar wind, and to search for the transition boundary between the solar wind and interstellar space as they head out of the solar system on escape trajectories. Both spacecraft are estimated to have about 20 years of useful life remaining.

Mission To Planet Earth

Planet Earth is in a constant state of change. Scientists understand some of the changes fairly well—weather in the short term, for example, or hurricane tracking, or how things grow. But they lack a lot of critical information, such as the kind of data needed to predict how the climate will shift a year hence and how the shift will affect farmers, water managers, fishermen and others whose livelihoods depend on climate.

That need is being met by a coordinated, U.S.-led international research program designed to reduce the uncertainties of global change. The United States has established the U.S. Global Change Research Program. NASA's part of the program is the Mission To Planet Earth (MTPE), a program that employs satellites and other tools to generate data about such areas of environmental concern as ozone depletion, deforestation, climate variability, earthquakes, volcanoes and destructive storms, and the ocean-influenced phenomenon known as El Niño.

MTPE studies are expected to yield improved weather forecasts, tools for managing forests and agriculture, information for fishing fleets and coastal planners, and—eventually—an ability to predict how the climate will change.

Phase I of the MTPE program has been under way since September 1991. Among the major contributors are the Upper Atmosphere Research Satellite, which is investigating the role of the upper atmosphere in climate and climatic change; the Shuttle-based Space Radar Laboratory, which has provided extensive data for studies of how shifting boundaries between temperate and boreal (northern) forests might affect climatic change; and the Total Ozone Mapping Spectrometer, an instrument flown on several U.S. and international spacecraft to study ozone depletion.

A prime contributor has been the U.S./French oceanographic satellite TOPEX/Poseidon, which precisely measures wave height and changes in average sea level. TOPEX/Poseidon reported rising sea levels over each of the past three years; though not sufficient to determine a trend, the findings are important to under-standing how the oceans interact with other Earth phenomena.

In a related discovery, TOPEX/Poseidon observations, combined with data from other sources, led to ground-breaking findings about the El Niño phenomenon. No one knows why El Niño occurs, but reports show it has been happening for hundreds of years, profoundly affecting weather patterns and causing floods and drought in various parts of the world. TOPEX/Poseidon data helped scientists predict El Niño more than a year in advance, enabling those potentially affected to alter crop plans and planning for other activities.

Additionally, NASA-developed weather satellites operated by the National Oceanic and Atmospheric Administration are contributing to a growing mosaic of Earth knowledge by

This is a visualization of the Antarctic ozone hole, computer-generated from data supplied by NASA's Total Ozone Mapping Spectrometer. The colors represent different ozone levels over the Antarctic; the purple area in the center is the "hole," an area where the ozone has disappeared.

Weather satellite imagery is contributing to a capability to predict weather events and initiate measures to prevent loss of life from destructive storms. This is an image supplied by the GOES-8 satellite of Hurricane Luis in September 1995; the white area shows the hurricane's high wind swirl as it moves over the Gulf of Mexico (blue) toward the Louisiana/Florida gulf coast (green).

providing an advanced observational/predictive capability that allows planning for upcoming weather events and initiation of measures to prevent loss of life from hurricanes and other destructive phenomena.

In 1997, MTPE will probe new areas with the launch of SeaStar, a privately-developed spacecraft designed to provide NASA with data on life in the oceans (data which may also be marketable to the fishing, oil and shipping industries), and with a joint NASA/Japan mission to study tropical rainfall, which is poorly understood but an essential element in the global climatic change equation.

(Continued)

The TOPEX/Poseidon oceanographic satellite has been a major contributor to NASA's Earth studies. It employs a sensitive altimeter to measure sea surface heights and compute average sea levels, information that helped enable advance prediction of the El Niño phenomenon.

Mission To Planet Earth *(Continued)*

The year 1998 will mark the beginning of Phase II of the Mission To Planet Earth program, which will feature the first integrated measurement of changes in global climate and will provide practical information to enhance the efficiency of business, farming, fishing and forestry operations.

A key tool of Phase II will be the Earth Observing System (EOS). Employing a number of different satellites, EOS will look at Earth in an entirely new way: rather than focusing on only one aspect (land, ocean, air), EOS satellites will observe multiple aspects of the planet. The broader range of information acquired will enable scientists to study interactions among Earth phenomena and allow them to move beyond a description of *what* is happening to and understanding of *why* it is happening.

EOS is the first system designed specifically to study the Earth as a complex series of interactions among life, air, water and land, the critical next step toward understanding and predicting the complexities of the global climate. EOS will contribute to such understanding by observing in 24 measurement areas which, a consensus of scientists believes, can supply answers to many of the complex questions about global climate change. Over a 15-18 year period, EOS satellites will fly over most of Earth's surface, gathering data on such matters as global climate changes in the atmosphere, land surface changes, pollution and water resources. What were previously studied as a series of isolated events will be examined as interconnected/interactive forces to form a snapshot of Earth as a whole.

The work of the EOS satellites will be complemented by a new family of spacecraft known as the Earth Systems Science Pathfinder (ESSP), a series of small, low-cost, rapid development science missions (life cycle cost will be capped at $120 million and development time will be held to 24-36 months). The ESSP satellites will

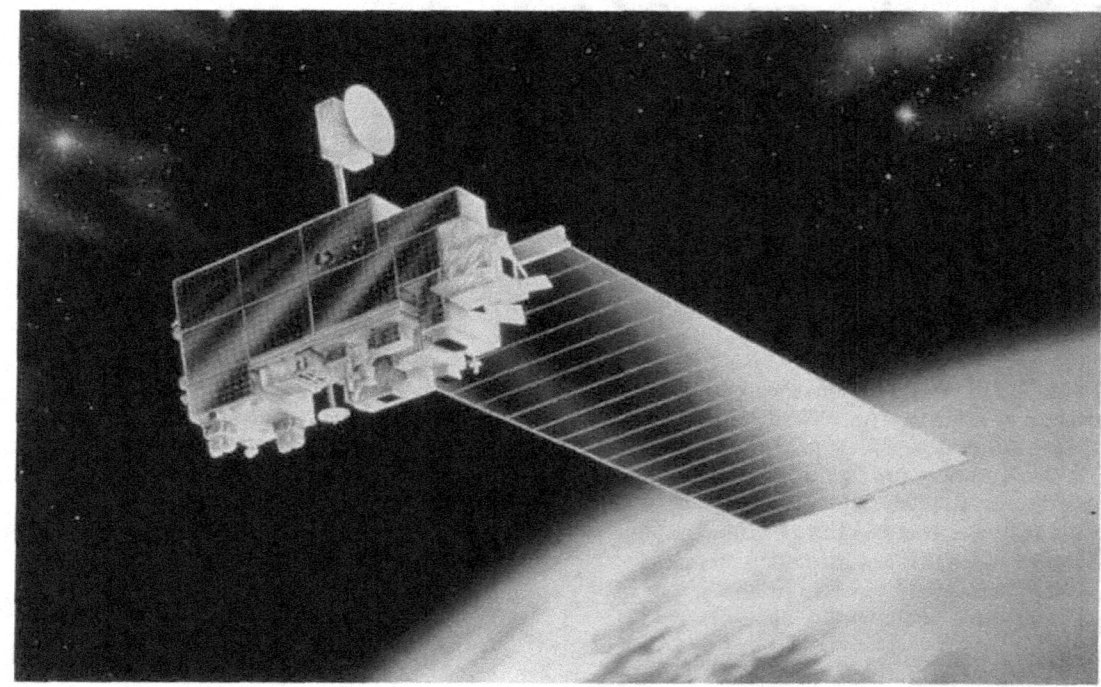

EOS AM-1, to be launched in 1998, will look at Earth in a new way: where earlier spacecraft observations focused on specific aspects of the global environment, EOS satellites will enable the first long term comprehensive measurements of how components of the Earth system interact.

incorporate the latest Earth observation technologies and will involve new contracting approaches and partnerships with industry. The first ESSP launch is targeted for 1999 and there will be one launch a year thereafter.

Another innovative NASA program, the New Millennium Program (NMP), will play a part in the Mission To Planet Earth. A coordinated NASA/industry activity incorporating next generation technologies, such as lightweight low-cost instruments, NMP focuses on demonstration of technologies and techniques for improving the performance and lowering the cost of future NASA missions. The first MTPE mission of this type will be the flight of EO-1, an advanced land imager to be launched in 1998. EO-2 is planned to follow 18-24 months later and there will be about one launch a year thereafter.

The MTPE program, which is generating an enormous and ever-increasing flow of data,

clearly requires a highly advanced data processing system to analyze and move data from the many U.S. and foreign spacecraft to governments, scientists, educators, businesses and the general public. For that purpose, NASA operates an information system known as EOSDIS (Earth Observing System Data and Information System). As the amount of data and the number of people using the system increase, EOSDIS will evolve and expand, taking advantage of the latest technological advances to maintain the quality and timeliness of information dissemination; thousands of users around the world will be able to access EOSDIS at the same time.

Slated for launch in 1998 is the EO-1 advanced land imaging spacecraft, the first Earth observation spacecraft of NASA's New Millennium Program, which features a revolutionary series of small, low-cost satellites. EO-1 will focus on demonstrating advanced technologies for future Earth observation missions.

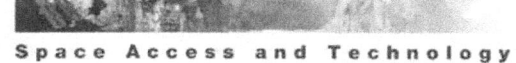

Technology for Space Research

NASA programs seek improved access to orbit, expanded commercialization of space systems, and technology advancement for enhanced U.S. competitiveness

A major NASA space research objective involves provision of innovative technology to enable ambitious future space missions and build capability for the U.S. space industry through focused technology efforts.

A milestone of that objective was a 1996 Space Shuttle mission devoted almost entirely to research directed toward expansion of the commercial space frontier. The mission was STS-77, Orbiter *Endeavour*, launched May 19; more than 90 percent of the payloads aboard *Endeavour* were sponsored by NASA's Office of Space Access and Technology (OSAT).

Primary payloads of STS-77 included the commercially-developed SPACEHAB module, which supported a wide range of commercial development experiments; an experiment in

deploying a large inflatable antenna; and a suite of four technology advancement experiments collectively known as TEAMS (Technology Experiments for Advancing Missions in Space).

Developed by SPACEHAB, Inc., Arlington, Virginia, the SPACEHAB Space Research Laboratory is intended to meet a need for additional experiment facilities on Shuttle flights. Carried in the Orbiter's payload bay and accessed through an airlock, SPACEHAB is a pressurized facility that roughly doubles the Orbiter's human-habitable volume and quadruples the volume available for crew-tended payload hardware. STS-77 carried the fourth flight of the module, SPACEHAB-04; it contained almost 3,000 pounds of flight hardware, virtually all of it in support of NASA's objective

to facilitate industry use of space for commercial products and services.

Two of the basic SPACEHAB modules were built and flight-certified in a development program financed entirely by private capital. Under contract to SPACEHAB, McDonnell Douglas Corporation conducted design, development and construction of the basic module. SPACEHAB also offers a double module, used on Shuttle/*Mir* missions, that can carry 6,000 pounds of cargo; it was modified by the Italian company Alenia Spazio.

SPACEHAB is operated as a commercial space system; the company pays NASA for launch services and leases experiment space to U.S. and international private industry, universities, research institutions and government agencies, including NASA.

On STS-77, *Endeavour* carried a single-module SPACEHAB in the forward portion of the Orbiter's payload bay; the module housed almost 3,000 pounds of experiments and support equipment for 10 commercial space development payloads. Some of these payloads were developed by NASA Commercial Space Centers, which are non-profit consortia of industry, academia and government partners dedicated to using the space environment to enable creation by industry of new and improved products and services.

Examples of the types of experiments carried out in the SPACEHAB laboratory include:

• The Advanced Separation Process for Organic Materials, designed to enhance separation technologies for medical products (sponsored by the Consortium for Materials Development in Space, University of Alabama-Huntsville).

• The Commercial Generic Bioprocessing Apparatus, a system for investigation of molecular, cellular, tissue, small animal and plant systems (sponsored by BioServe Space Technologies, NASA's Commercial Space Center at the University of Colorado, Boulder).

• A series of Commercial Float Zone Furnace experiments designed to produce large, ultrapure compound semiconductor and mixed oxide crystals for electronic devices and infrared detectors (the experiments are joint efforts of Marshall Space Flight Center, the Canadian Space Agency and the German Space Agency).

(Continued)

Carried in the Space Shuttle Orbiter's payload bay, the SPACEHAB space research laboratory doubles the habitable volume of the Orbiter and quadruples the volume available for crew-tended experiment hardware.

Space Technology Development *(Continued)*

On Day Two of the STS-77 mission, the *Endeavour* crew deployed the first of two satellites to be released from the Orbiter, to conduct a major event of the 10-day flight; the Inflatable Antenna Experiment (IAE).

The IAE was designed to lay the groundwork for future technology developments in inflatable space structures, which have potential to be 10 to 100 times less expensive than conventional structures. The experiment was carried aboard the 1,866-pound Spartan 207, a multipurpose free-flying satellite that is deployed from the Orbiter and retrieved by an astronaut operating the Shuttle's Remote Manipulator System. Managed by Goddard Space Flight Center, the carrier version of the Spartan was making its second flight; overall it was the eighth Spartan mission flown on the Space Shuttle.

Packed into the Spartan, the IAE weighed only 132 pounds; on deployment (inflation), it would expand to an antenna 50-feet in diameter mounted on three 92-foot struts. The IAE was developed by L'Garde Inc., Tustin, California and Jet Propulsion Laboratory under NASA's In-Space Technology Experiment Program.

Spartan was released on May 20; the antenna was successfully deployed and it achieved the proper configuration. The inflation process was captured by the STS-77 crew on still, motion picture and video cameras. For post-mission analysis of the inflatable structure's performance, the antenna surface was illuminated by arrays of lights mounted on the Spartan satellite and the resulting patterns were acquired by Spartan's video recorders. After 90 minutes of operation, the IAE was jettisoned;

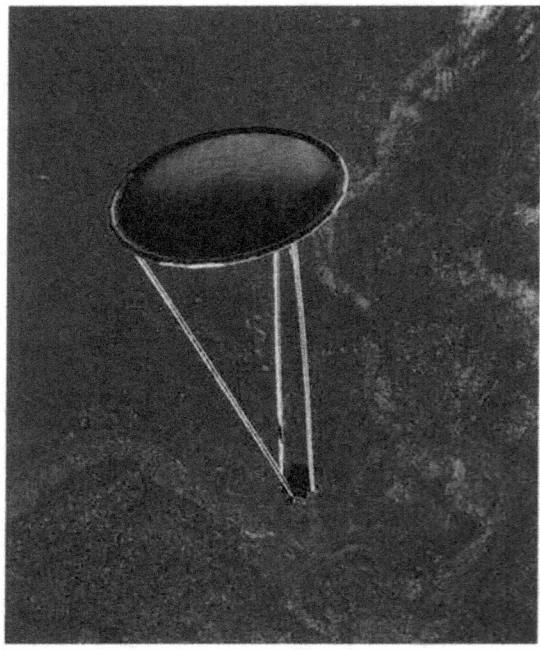

Pictured against the backdrop of St. Louis and the Mississippi River, this large antenna was released by the free-flying Spartan 207 satellite in an STS-77 investigation of the potential of inflatable space structures. It was fitted into Spartan as a compact 132-pound package, then inflated into a 50-foot-diameter, 92-feet-long structure.

This view of the Orbiter Endeavour's stern shows the Spartan 207 free flyer being recaptured after a day-long research trip away from the Orbiter. It is being jockeyed toward its stowage berth by an astronaut operating the remote manipulator system.

the Spartan was grappled and retrieved the following day.

On May 23, Day Four, the *Endeavour* crew deployed the second satellite, this one known as PAMS, for Passive, Aerodynamically-stabilized Magnetically-damped Satellite. The PAMS experiment consisted of the small (115 pounds) satellite and a measuring system that enabled the crew to observe the satellite's motions after deployment from the Orbiter. The experiment was a demonstration of aerodynamic stabilization, a technique that can be used to position a satellite in a specific orientation while in low Earth orbit.

The PAMS satellite was spring-ejected from *Endeavour's* payload bay as cameras in the Orbiter recorded the deployment. For most of the next five days, the Orbiter trailed the 20-

STS-77 astronauts Marc Garneau (foreground) and Curtis L. Brown, Jr. check out the audio system of the SPACEHAB Space Research Laboratory. SPACEHAB is a commercially-developed pressurized facility carried in the Shuttle Orbiter's payload bay.

inch satellite at a distance of about 2,000 feet while the Shuttle crew used the measurement system to note the damping (stabilization) motions of the satellite. Video and radar data were acquired throughout the station-keeping period.

PAMS was one of four experiments in the TEAMS group of payloads mounted in Goddard Space Flight Center's Hitchhiker experiment carrier in *Endeavour's* cargo bay. The others were:

• The Global Positioning System (GPS) Attitude and Navigation Experiment, a test of how accurately the GPS constellation of positioning satellites can determine the attitude of a vehicle in an orbital environment. The International Space Station will use GPS for attitude determination as well as velocity and time information.

• The Vented Tank Resupply Experiment, which tested improved methods for in-space refueling to provide data for future designs of spacecraft liquid fuel storage tanks. Lewis Research Center and contractor Lockheed Martin developed the experiment.

• The Liquid Metal Thermal Experiment, an evaluation of the performance of liquid metal heat pipes in microgravity conditions. Heat pipes are thermal management devices used on many existing spacecraft for waste heat removal. However, the operational characteristics of heat pipes are not completely understood, because they have not been operated at high temperatures in microgravity. The three heat pipes in this experiment contained potassium and operated at very high temperatures, 300 to 1000 degrees Celsius. The data obtained will be invaluable to designers of space systems requiring high temperature heat rejection.

STS-77, an eventful and highly productive space technology development mission, ended on May 29 when *Endeavour* touched down at Kennedy Space Center after a flight of 10 days and 39 minutes.

Reusable Launch Vehicle

NASA's Reusable Launch Vehicle (RLV) program has dual objectives: to demonstrate technologies leading to a new generation of space boosters capable of delivering payloads at significantly lower cost, and to provide a technology base for development of advanced commercial launch systems that will make U.S. aerospace manufacturers more competitive in the global market.

Launched in 1994, the RLV program moved ahead on two fronts in 1996 with a restructuring of the X-34 air-launched small booster project and selection of a contractor for development of the larger X-33 technology demonstrator (see page 30).

On June 10, 1996, NASA announced a program for the X-34 that will be a technology demon-

Under development by Orbital Sciences Corporation is the X-34, an experimental vehicle for testing technologies for a single-stage-to-orbit launch vehicle. First flight is targeted for 1998.

strator rather than an operational system. Orbital Sciences Corporation (OSC), Dulles, Virginia was selected as prime contractor from among nine bidders.

The X-34 will weigh about 45,000 pounds and will be air-launched from a Lockheed L-1011 jetliner. The craft will be about 58 feet long and span about 28 feet.

The aim of the revised program is to create a reusable, suborbital test vehicle for demonstrating single-stage-to-orbit (SSTO) technologies, most of which have applicability in the X-33 program.

The X-34 plan calls for development of a vehicle capable of attaining a velocity of Mach 8 (roughly 5200 miles per hour), flying to an altitude of about 50 miles, and returning to an Earth base for a wheeled landing. The initial test flight must be made by October 1, 1998.

Among the technology areas the X-34 will explore is a very high flight rate: the vehicle must demonstrate potential for 25 flights a year. Other areas to be investigated are advanced thermal protection systems, a composite airframe, reusable propellant tanks, autonomous landing systems and advanced avionics.

OSC's development effort will be backed by a strong NASA/industry team. The X-34's propulsion system will be provided by Marshall Space Flight Center, which has been working since 1994 on an advanced liquid oxygen/kerosene rocket engine with a low-cost, partially reusable thrust chamber. The X-34 engine will have 60,000 pounds of thrust; since there is no requirement for attaining orbit in the test program, no second stage is needed.

Ames Research Center will contribute thermal protection technology and Langley Research Center will handle computational fluid dynamics and aerodynamic testing.

Industry firms participating include AlliedSignal Aerospace, Torrance, California, which is developing advanced avionics and hydraulic power systems; Oceaneering Space Systems, Houston, Texas (re-entry thermal control); and Charles Stark Draper Laboratory, Cambridge, Massachusetts (navigation and guidance technology).

(Continued)

Reusable Launch Vehicle *(Continued)*

On July 2, 1996, Vice President Albert Gore announced the selection of Lockheed Martin Corporation for Phase II development of an experimental space booster that could become the prototype of a 21st century operational reusable launch vehicle (RLV).

The vehicle is the X-33 Advanced Technology Demonstrator, a half-scale version of the proposed operational system which, Lockheed Martin says, will be able to deliver 40,000 pounds of payload to low Earth orbit at a small fraction of the current cost of launching payloads to orbit. Under a cooperative agreement, the X-33 will be developed jointly by NASA and Lockheed Martin's famed "Skunk Works," noted for its success in creating such radical aerospace craft as the U-2 reconnaissance plane, the SR-71 Mach 3 research plane, and the F-117 Air Force stealth fighter.

The aim of the RLV program, NASA Administrator Daniel S. Goldin said at the announcement ceremony, is "to build a vehicle that takes

In development for first flight in 1999 is the NASA/Lockheed Martin X-33 Advanced Technology Demonstrator, intended as the prototype of a 21st century low-cost-to-orbit reusable launch vehicle.

days, not months, to turn around; dozens, not thousands, of people to operate; with launch costs that are a tenth of what they are now. Our goal is a reusable launch vehicle that will cut the cost of getting a pound of payload to orbit from $10,000 to $1,000."

Goldin underlined the fact that the RLV project represents a "radical departure" from the way NASA has done business in the past. In line with its mission of financing high risk developments in the interests of U.S. competitiveness and the national economy, NASA is funding the major share of the X-33 development, but the 21st century operational RLV will be developed on a commercial basis and NASA will become a user, rather than an operator, of the system.

NASA is providing $941 million through 1999 for the X-33; Lockheed Martin and industry partners will invest some $220 million in startup costs and initial R&D. The development will be carried out under a cooperative agreement—a true government/industry partnership—that is different from conventional contracts; it is a performance-based agreement under which NASA will make progress payments only when the industry team completes predetermined milestones.

The Phase II award—for design, construction and flight testing of the X-33—follows a 15-month Phase I concept definition effort during which each of three contractors developed its own design, operations plan and business investment strategy. After a lengthy evaluation of all the bids, Lockheed Martin was selected as NASA's industry partner over competitors McDonnell Douglas Corporation and Rockwell International.

Known as Venture-Star, the Lockheed Martin design is a wedge-shaped lifting body type of vehicle in which the entire airframe, not just the stubby winglets, generates lift. It is designed to be launched vertically and land horizontally like the Space Shuttle, but where the Shuttle

uses conventional engines, a large throwaway fuel tank and drop-off boosters for launch, the X-33 will be totally reusable and employ what are known as "aerospike" engines. Built by Rockwell Rocketdyne, the engines offer utmost efficiency through a system that does not use conventional bell-shaped engine nozzles but automatically compensates for decreasing atmospheric pressure and regulates thrust as the vehicle ascends.

Lockheed Martin's industry partners include AlliedSignal, Teterboro, New Jersey (subsystems, avionics and operations support); Rockwell Rocketdyne Division, Canoga Park, California (aerospike propulsion); Rohr, Inc., Chula Vista, California (thermal protection system); Sverdrup Corporation, St. Louis, Missouri (launch site architecture and engineering); and Alliant Techsystems, Magna, Utah (fuel tanks).

The X-33 is being developed in a "fast track" program that envisions first flight in March 1999. An incentive clause provides a bonus if the Venture-Star can make 15 flights by the end of the century, at least one of them to a velocity of Mach 15. A successful X-33 program could lead to commercial development of a full-scale RLV and first operational use around 2005.

Advanced Concepts Research

NASA has initiated a program of studies and experiments intended to identify and define advanced concepts that have potential for affecting revolutionary improvements in future U.S. space activities. A key element of this program is the Advanced Concepts Research Program (ACRP).

Sponsored by the Office of Space Access and Technology, this effort addresses diverse, highly innovative technical concepts, such as electromagnetic catapults, beamed energy, small spacecraft swarms, propellantless propulsion by means of tethers, ultra-large telescopes and a broad array of concepts.

Eight innovative proposals were selected in February 1996 from more than 100 proposals submitted in response to the initial solicitation.

ACRP projects are led by principal investigators designated ACRP Fellows and are funded to a maximum amount of $250,000. While the

Fellows' research will focus on their own proposed concepts and technologies, each Fellow will also serve as a member of a broad interdisciplinary advanced concepts team.

The ACRP program contemplates selection of at least eight new ACRP projects each year. Each project will be conducted over a 24-month period, so there will be approximately 16 ongoing projects in any year. Interactions among ACRP Fellows and NASA researchers will be accomplished through workshops and periodic meetings, such as a "virtual research center" created by use of the Internet, and through pioneering infrastructures.

Infrared Camera

Teaming with an industry partner, NASA has developed a revolutionary infrared camera that offers important applications not only in aerospace research but in such other areas as air transportation, environment monitoring and medicine.

The camera was developed by the Center for Space Microelectronics Technology at Jet Propulsion Laboratory, Santa Barbara, California, in cooperation with Amber, a Raytheon company, Goleta, Georgia.

An innovative feature of the camera is its use of highly sensitive quantum-well photodetectors or QWIPS. The greater sensitivity of long wavelength QWIPS could allow physicians to detect tumors using thermographic (heat analysis) techniques; improve pilots' night vision to allow better landings; and enable environmental scientists to monitor pollution and weather patterns with enhanced measurement accuracy. Other possible applications include law enforcement, industrial process control, search and rescue, and military antimissile surveillance.

The camera weighs only 9.9 pounds and measures 4.4 inches wide, 10.3 inches deep and 7.2 inches long. The prototype plugs into a wall socket for power but the camera can readily be converted to battery power for portability.

Because infrared light detectors must operate at extremely low temperatures, the camera contains a Stirling cryocooler, a closed-cycle refrigerator about the size of a fist that cools the camera from room temperature to about 343 degrees below zero Fahrenheit in about 10 minutes. The QWIPS technology represents a half decade of development effort on the part of the Center for Space Microelectronics Technology and its industrial affiliate under contract to NASA's Office of Space Access and Technology.

An industry/government team developed this revolutionary infrared camera that has broad applications in medicine, environment monitoring, industrial processing and law enforcement, as well as in aerospace research.

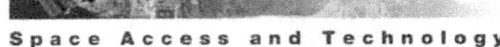
Small Satellite Developments

NASA's Small Spacecraft Technology Initiative (SSTI) is a program designed to demonstrate technologies for reducing the cost and the time of getting civil and commercial space missions from the drawing board to orbit. The program features new approaches to satellite design and development that will not only significantly lower cost but will also permit the builder to incorporate commercial standards in the design and qualification process. Thus, the program will serve NASA's needs by allowing more frequent space missions within predictably lower budgets, and it will additionally enhance the competitive posture of U.S. space system manufacturers in the global market.

The program was launched in 1994 with the awards of contracts to two industry teams for development of "smallsats" or "lightsats" in the 600-900 pound class. The small satellites are designed to accommodate a wide range of missions through use of standard hardware and software adapted to various applications. The initial satellites are named "Clark" and "Lewis" for the leaders of the 19th century U.S. expedition to the Pacific northwest. Although the expedition is invariably called "Lewis and Clark," the order is reversed for the NASA program because Clark was scheduled for the first launch. At *Spinoff* publication time in midyear 1996, Clark was targeted for November launch; Lewis was to follow at yearend or early in 1997.

Clark is being developed by CTA, Incorporated, Rockville, Maryland with Lockheed Martin Astronautics as principal subcontractor. CTA is among the world leaders in manufacturing small space systems, having built, launched and operated 21 lightsats, with several others in development.

Clark is intended to demonstrate 36 advanced technologies, among them image data compression, a mini star tracker, a low-cost Sun sensor, advanced composite structures, room tempera-

Developed by CTA, Incorporated, the Clark satellite is designed to demonstrate technologies for a new family of satellites featuring reduced cost and compressed development time.

ture x-ray detectors, 3D imaging of atmospheric trace gases, and on-board data processing.

Although essentially a demonstrator of SSTI concepts, Clark will also operate as a science and applications satellite with three science payloads and a remote sensing imaging system. Principal focus will be on commercial remote sensing which, CTA believes, has promise of becoming a multibillion dollar industry. Science research includes x-ray spectrometry, including atmospheric pollution measurements, cloud detection and atmospheric tomography.

The Lewis satellite is being developed under the team leadership of TRW Inc., Redondo Beach, California. One of the nation's largest aerospace manufacturers, TRW developed smallsat technology for a number of military programs and is currently building commercial lightsats. Like Clark, Lewis is a technology demonstrator but it will also carry science

payloads and it will operate as a pathfinder to stimulate the commercial market.

Lewis incorporates more than 25 new spacecraft and payload technologies designed to provide superior performance while minimizing cost and schedule time, among them miniaturized cryocoolers, advanced composites, faster data processors, lightweight propellant tanks and smaller star trackers.

A key feature of the Lewis spacecraft is a new Hyperspectral Imager that will generate Earth imagery in 384 spectral bands (the NASA-developed Landsat satellite has seven bands). The greater capability of the imager will allow analysts to distinguish among a much broader range of Earth features. The satellite will also carry an extreme ultraviolet spectrometer, built by the University of California at Berkeley, to take images of the night sky and cosmic background.

Clark's sister satellite—Lewis—has an advanced remote sensing imager that significantly broadens the range of Earth features that can be analyzed. Both satellites will operate as commercial remote sensing systems in addition to their NASA assignments as technology demonstrators.

Toward Future Flight

NASA's aeronautical research program is providing a technology base for tomorrow's safer, more efficient, more competitive aircraft

On March 19, 1996, NASA and industry partner McDonnell Douglas Corporation (MDC) unveiled to the public a new subsonic flight vehicle designated X-36. A remotely-piloted tailless research craft, the X-36 is designed to demonstrate the feasibility of future tailless military fighters that can achieve agility levels superior to those of today's aircraft.

In the absence of a tail, control of the X-36 is accomplished by a combination of thrust vectoring (maneuvering by directing the engine's exhaust flow) and innovative aerodynamic control features. Tailless fighter configurations offer reduced weight, increased range and improvement in survivability; the X-36 program is intended to establish confidence to incorporate these technologies in future piloted vehicles.

The unmanned X-36 is "flown" by a pilot located in a van at the flight test facility; a camera in the X-36 cockpit relays instrument readings and displays to a console in the van. With a wing span of only 10.4 feet and a gross weight under 1,300 pounds, the X-36 is powered by a single turbofan originally designed as a cruise missile power plant.

The subscale vehicle was selected for affordability, in line with NASA's "better, cheaper, faster" approach to new aerospace developments. At 28 percent scale, it enables demonstration of all key control integration technologies at a fraction of the cost of a full-scale piloted aircraft.

The initial X-36 was developed and built in only 28 months; a second model was under construction at midyear 1996. The program

resulted from a 1994 cost-sharing NASA/MDC agreement under which Ames Research Center is responsible for continued development of critical technologies and MDC is responsible for fabrication of the two X-planes.

At *Spinoff* publication time, the X-36 was being readied for a midsummer start of the flight test program at Dryden Flight Test Center. The combined cost for development, fabrication and flight testing is approximately $17 million.

The X-36 project exemplifies one aspect of a broad NASA aeronautical research and technology program that seeks to improve the performance, efficiency and environmental characteristics of all types of planes, and additionally addresses such infrastructure factors as air traffic control, navigation and communications.

Basic research of a general nature aims at advancing aerodynamics, propulsion, materials and structures, aviation electronics, and knowledge of the human factors in flight operations. Another part of the program embraces technology development for specific types of flight vehicles, such as high performance military aircraft or the tiltrotor type of transport on the near horizon. A third part of the program seeks solution of current and predictable aviation problems, such as reducing airplane and helicopter noise levels, finding ways to alleviate air traffic congestion, and a variety of safety-related investigations.

Among priority objectives are development of payoff technologies for a new generation of economic, environmentally acceptable U.S. subsonic aircraft and a safe, highly productive air transportation system; building a technology base for an economically viable second generation supersonic passenger transport; developing and demonstrating technologies for airbreathing hypersonic flight; and maintaining/operating critical facilities for aeronautical research in support of industry and technology-generating government agencies.

NASA pursues these objectives through in-house research and cooperative endeavors with academia, industry and other government agencies. NASA's principal aeronautical research facilities are Ames Research Center, Moffett Field, California; Langley Research Center, Hampton, Virginia; Lewis Research Center, Cleveland, Ohio; and Dryden Flight Research Center, Edwards, California. Examples of their activities are described on the following pages.

Designed jointly by NASA and McDonnell Douglas Corporation, the X-36 is a subscale, remotely-piloted tailless vehicle for demonstrating technologies that could lead to lighter, longer-ranging, more survivable, more agile military fighter aircraft.

High Speed Research

Aircraft manufacturers of several nations are developing technology for the next plateau of international aviation competition: the long-range, environmentally-acceptable second generation supersonic passenger transport, which could be flying by 2010.

Predicting large-scale increases in demand for long-haul overwater passenger transportation early in the next century, market experts see a need for some 500 next generation supersonic transports worth an estimated $200 billion and 140,000 jobs.

Capturing a major share of that market is vitally important to a U.S. aerospace industry that is transitioning from a traditionally defense-dominated product line to a commercially driven manufacturing activity. To help boost the industry's competitiveness, NASA is conducting a High Speed Research (HSR) program that addresses the highest priority, highest risk technologies for a High Speed Civil Transport (HSCT). The HSR program is intended to demonstrate the technical feasibility of the vehicle; the decision to proceed with full-scale development will be up to industry.

This McDonnell Douglas conceptual design for a Mach 2.4 (1600 miles per hour) supersonic transport is sized to carry about 300 passengers over a distance of 5,000 nautical miles. A NASA/industry High Speed Civil Transport research effort is a first step toward determining whether such a plane can be economically viable and environmentally acceptable.

The program is being conducted as a national team effort with shared government/industry funding and responsibilities. The team includes NASA's Langley, Lewis and Ames Research Centers and Dryden Flight Research Center; engine manufacturers GE Aircraft Engines and Pratt & Whitney division of United Technologies; airframe manufacturers The Boeing Company, McDonnell Douglas Corporation and Rockwell North American Aircraft Division; other manufacturers; materials suppliers; and academic institutions.

The team has established a baseline design concept that serves as a common configuration for investigations. A full-scale craft of this design would have a maximum cruise speed of Mach 2.4, or about 1600 miles per hour, only marginally faster than the currently operational Anglo-French Concorde supersonic transport. However, the HSCT would have about double the range and triple the passenger capacity of the Concorde, and it would operate at an affordable ticket price, estimated at 20 percent above comparable subsonic flight fares.

Phase I of the HSR program, which began in 1990 and continued through 1995, focused on environmental challenges: engine emission effects on the atmosphere, airport noise and the sonic boom. Much research remains to be accomplished in these and other areas, but Phase I established some clear lines of approach to major problems and spawned confidence among team members that environmental concerns can be satisfied.

Phase II, initiated in 1994, focuses on the technology advances needed for economic viability, principally weight reductions in every aspect of the baseline configuration, because weight affects not only the aircraft's performance but its acquisition cost, operating costs and environmental compatibility. In materials and structures, the HSR team is developing, analyzing and verifying the technology for trimming the baseline airframe by 30-40

percent; in aerodynamics, a major goal is to minimize air drag to enable a substantial increase in range; propulsion research looks for environment-related and general efficiency improvements in critical engine components, such as inlet systems. Phase II includes computational and wind tunnel analyses of the baseline HSCT and alternative designs. Other research involves ground and flight simulations aimed at development of advanced control systems, flight deck instrumentation and displays.

In 1996, the HSR program moved beyond laboratory investigations into the actual supersonic flight realm through a NASA agreement with the Russian Tupolev Design Bureau, developers of the first supersonic transport, the TU-144, which first flew in passenger service in 1977. Under the agreement, a modified TU-144LL supersonic flying laboratory is providing up-to-date information of "real world" conditions in which the next generation supersonic transport will fly. The TU-144LL rolled out of its hangar on March 17 to begin a six-month, 32 flight test program.

(Continued)

Shown at a March 1996 rollout ceremony, the Russian TU-144LL supersonic flying laboratory is participating in NASA's High Speed Civil Transport research program.

High Speed Research *(Continued)*

The TU-144LL can fly at Mach 2.3, or about 1500 miles per hour, close to the speed of the HSCT baseline concept (Mach 2.4) and is thus an ideal vehicle for NASA studies of high temperature materials and structures, acoustics, supersonic aerodynamics and supersonic propulsion.

The TU-144LL is one of 17 TU-144s built. The major modification for the HSR work is a change of engines. The original engines were replaced by newer and larger NK-321 augmented turbofans initially employed to power Tupolev's TU-160 Blackjack bomber. Among a number of other upgrades and modifications, the jetliner's passenger seats were removed to make room for the six NASA/U.S. industry experiments' instrumentation and data collection systems. Two additional experiments are to be conducted on the ground using a TU-144 engine.

The flight deck portion of the HSR program also progressed to flight status in 1996 with a series of tests to investigate a "synthetic vision" concept that could obviate the need for forward-facing cockpit windows. The reason for this departure from conventional design philosophy is the fact that a supersonic transport of the baseline configuration would land nose-high—as do the Concorde and the TU-144—with the flight deck 45 feet above the runway and more than 50 feet forward of the landing gear. In that position, the pilots have no view of the runway ahead of them.

In the first generation supersonic transports—the Concorde and the TU-144—the forward vision problem was solved by use of a mechanism that lowers—or "droops"—the forward part of the nose section for takeoffs and landings and thereby affords a clear view forward. The mechanism, however, imposes a

The Russian TU-144LL supersonic flight laboratory employs a mechanical system to "droop" the nose section. This technique is necessitated by the fact that the airplane lands nose high and pilots could not see the runway with the nose in standard flight position. The NASA/industry High Speed Research team is working on an alternative approach (see photo opposite).

heavy weight penalty that is not considered acceptable for the second generation vehicle.

A potential solution devised by the HSR team is the external visibility system (EVS), a group of sensors and imaging systems that would feed large-format cockpit displays of high resolution imagery and computer graphics. The EVS could eliminate forward-looking cockpit windows and obviate the need for the heavy, expensive mechanical nose-drooping system.

In the second generation supersonic transport, the EVS could save thousands of pounds of droop mechanism weight, weight that could be used to allow increased passenger capacity or greater range. The synthetic vision system might also find utility in subsonic air transportation, allowing pilots to fly and land safely in low visibility conditions; that would enable increasing the number of flights in poor

Future jetliners may employ a design technique that eliminates forward-facing cockpit windows and substitutes a 3D computer-generated color display to give the pilots "synthetic vision" on takeoffs and landings. Already flight tested, this system could save thousands of pounds of weight that could be more productively used.

weather, reducing terminal delays and cutting costs for airlines and passengers.

The HSR synthetic vision system was tested in a series of flights in 1995-96 at NASA's Wallops (Virginia) Flight Facility and at Langley Air Force Base in Hampton, Virginia. Sensors tested included a digital video camera, three infrared cameras and two microwave radar systems. The tests were flown on Langley Research Center's Transport Systems Research Vehicle (TSRV), a Boeing 737 equipped with a windowless research cockpit in the passenger section in addition to the normal windowed cockpit, and in a Westinghouse BAC 1-11 avionics test aircraft.

The flight test program consisted of two phases. During the sensor data collection phase, the TSRV and the BAC 1-11 flew typical approach, cruise and holding patterns, testing the capability of the sensors to detect airborne traffic and ground objects. During the pilot-in-the-loop phase, the TSRV flew approaches and landings controlled from the research cockpit and tested the pilots' ability to control and land the aircraft relying only on sensor/computer-generated images and symbology.

All planned in-flight test points were achieved, and extensive data was collected from the radar, infrared and video sensors. More than 80 windowless piloted approaches and landings were successfully conducted by pilots from Langley and Ames Research Centers, Boeing and McDonnell Douglas. Initial pilot comments and performance reports were encouraging with respect to the feasibility of using sensor/symbology displays for flight path control.

In addition to the principal members of the HSR team, the flight deck research included Honeywell, Inc., Phoenix, Arizona; Rockwell Collins, Cedar Rapids, Iowa; FLIR Systems, Portland, Oregon; and Westinghouse Electric Corporation.

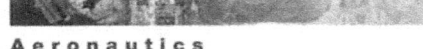
Hypersonic Research

An important part of NASA's aeronautical research involves development of technology for future vehicles capable of airbreathing hypersonic flight, which by definition is flight faster than Mach 5, or five times the speed of sound (about 3300 miles per hour).

Over the past decade, NASA has conducted several hypersonic exploratory investigations and built a technology base in such areas as hypersonic propulsion, cryogenic fuel, materials and structures, computer science, and integration of engines with hypersonic airframes. A technology of particular interest is the scramjet (supersonic combustion ramjet) propulsion system, which burns a mix of hydrogen carried aboard the vehicle and oxygen scooped up from the atmosphere.

In 1997, NASA will initiate a new effort to advance hypersonic technology for airbreathing vehicles. Called Hypersonic X-Vehicle (Hyper-X), it is a technology development program intended to further expand the knowledge base in this area of aeronautics. Successful development of the essential technologies would afford potential for future application in global reach military aircraft or, in space operations, as the first stage of a two-stage-to-orbit vehicle.

Preliminary design, analysis and wind tunnel work was conducted in 1995-96. The program contemplates a five-year effort, beginning in Fiscal Year 1997, involving wind tunnel and flight tests of a subscale, unmanned experimental airframe powered by an airframe-embedded scramjet engine. Flight tests, to begin in 1998, will be conducted at specified test points up to Mach 10.

One possible configuration of a future airbreathing hypersonic vehicle that could cruise within the atmosphere at speeds up to Mach 10.

Winter Runway Safety Study

In cooperation with the U.S. Federal Aviation Administration (FAA) and Transport Canada, NASA has initiated a new study of winter runway friction. Langley Research Center manages the NASA portion of the program. Several European aviation organizations and equipment manufacturers are also participating.

The five-year government/industry study, called the Joint Winter Runway Friction Measurement Program, includes braking tests with instrumented aircraft and ground vehicles in the U.S. and Canada. The results are expected to enhance safety for all ground operations and help relieve airport congestion during bad weather. Additionally, the research will help industry develop improved tire designs, better chemical treatments for snow and ice, and runway surfaces that minimize bad weather effects.

A near-term program goal is improved flight crew recognition of less-than-acceptable runway friction conditions prior to the "go/no go" and "land/go around" decision points.

An initial set of tests was conducted at Jack Garland Airport, North Bay, Ontario using NASA's Boeing 737 Transport Systems Research Vehicle (TSRV) and a Canadian National Research Council Falcon-20. Surface conditions were artificially varied to expand the range of data collected. Many different friction measuring ground vehicles—vans, trailers and modified cars—took readings with continuous and fixed slip devices under similar runway conditions for comparison with each other and with the braking performance of the instrumented aircraft. Further evaluations were planned at Brunswick Naval Air Station, Maine; water contamination tests were scheduled at NASA's Wallops (Virginia) Flight Facility and the FAA Technical Center.

Data from the program will be used to qualify the degree of improvement in measuring runway friction since NASA and FAA teamed on similar tests in the mid-1980s. There is need to evaluate improved measurement equipment, software and test procedures developed since the earlier tests, and there is need for data on new anti-icing and de-icing chemicals, water/slush drag effects on new aircraft, and tire construction effects on hydroplaning.

In a spinoff application, much of the equipment being used to monitor runways is being—or will be—used to measure highway pavement friction performance. In areas with high accident rates, pavement textures can be modified, on the basis of friction measurements, to improve the safety of auto travel.

NASA, the Federal Aviation Administration and Transport Canada are teaming on a five-year winter runway friction investigation to enhance airport ground safety. Here NASA's instrumented 737 test aircraft is landing on a snow-covered runway at North Bay, Ontario.

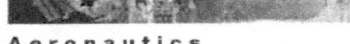
Flight Research

In March 1996, NASA initiated flight testing of a new thrust vectoring concept that could lead to significant increases in the performance of both civil and military aircraft flying at subsonic or supersonic speed.

The tests at Dryden Flight Research Center are part of a program known as ACTIVE (Advanced Controls Technology for Integrated Aircraft), a collaborative effort of NASA, the Air Force's Wright Laboratory, McDonnell Douglas Aerospace (MDA), and Pratt & Whitney Government Engines & Space Propulsion unit (P&W).

The test aircraft is a twin-engine F-15 ACTIVE, a modified version of the Air Force F-15B fighter built by MDA and powered by F-100-PW-229 engines, each of which is equipped with a nozzle that can swivel 20 degrees in any direction, giving the aircraft thrust control in the pitch (up and down) and yaw (left-right) directions. This vectored (deflected) thrust system could replace conventional drag-inducing aerodynamic controls and thereby gain increased fuel economy or range.

The tests began with four flights in March/April, then progressed to the first supersonic flight on April 24. On that occasion, the F-15 ACTIVE successfully demonstrated both pitch and yaw deflections at speeds of Mach 1.2 to 1.5. The flight test plan contemplated about 60 flights totaling 100 hours at speeds up to Mach 1.85 and angles of attack (the angle between the aircraft's body/wings and its actual flight path) up to 30 degrees.

The F-15 ACTIVE program is representative of the type of flight research conducted by NASA

The F-15 ACTIVE is exploring a vectored thrust system that could replace conventional aerodynamic control surfaces.

to explore new technologies and new flight regimes. NASA conducts such programs independently or in cooperation with U.S. industry and the Department of Defense, sometimes in cooperation with international development teams.

Another example of a Dryden flight research program is NASA's High Alpha investigation. High Alpha refers to high angles of attack, a flight regime in which the airflow becomes extremely complex. To provide aircraft manufacturers with a technology base for designing high performance aircraft capable of "supermaneuverability" and of maintaining stability/controllability at high angles of attack, NASA conducted the decade-long High Alpha program that concluded on May 29, 1996 with the final flight of NASA's F-18 HARV (High Alpha Research Vehicle).

In the first phase of the program, initiated in 1987, the F-18 HARV explored angles of attack up to 55 degrees. In the second phase, NASA investigated thrust vectoring technology to determine the impact on aircraft maneuverability at high angles of attack. In the final phase, the F-18 HARV's handling qualities were evaluated by 14 different pilots representing NASA, the Department of Defense, and support contractors McDonnell Douglas Aerospace and Calspan Corporation.

Among other flight projects under way at Dryden are two examples of test programs intended to support NASA activities not directly connected with aeronautics advancement. One is a project involving airborne tests of an advanced thermal protection system (TPS) for use on the X-33 Reusable Launch Vehicle (see page 30). The project employs an F-15B Flight Test Fixture-II (FTF-II) aircraft for atmospheric testing (the ascent and landing phases of the launch vehicle's operation), where the potential threat to the TPS is impact with rain drops, cloud droplets or ice crystals. Test participants include Marshall Space Flight Center and Rockwell International.

Another new program involves testing the Theseus, a robot aircraft to be employed in NASA's Mission To Planet Earth program for research in such areas as stratospheric ozone depletion and the atmospheric effects of future high speed civil transport engines. Built by Aurora Flight Sciences Corporation, Fairmont, West Virginia, Theseus is a twin-engine propeller-driven craft with a 143-foot wingspan. Constructed largely of composite materials, it is capable of carrying 700 pounds of science instruments to altitudes above 60,000 feet for durations of more than 24 hours. The plane made its initial flight at Dryden in May 1996.

NASA's veteran F-18 HARV research aircraft concluded a decade-long flight test program that explored aircraft maneuverability and controllability at high angles of attack.

Technology Twice Used

A representative selection of new products and processes adapted from technology originally developed for NASA mainline programs, underlining the broad diversity of spinoff applications and the social/economic benefits they provide

A Boon for Bone Research

Among spinoffs in the field of health and medicine is a unique device for analyzing bone tissue

Astronauts operating in a weightless environment for long periods are subject to a form of bone deterioration known as disuse osteoporosis, which can pose a serious hazard for crews on future space missions that might run two years or more. After long exposure to microgravity, weight bearing bones lose calcium and density, become very brittle and are easily fractured.

For two decades, NASA has been engaged in study and experimentation toward developing countermeasures—special nutritional/exercise programs, for example—to microgravity-induced bone deterioration. A particular need of this research was a way of making direct measurements of bone stiffness and mass; while there were available instruments for determining the mineral density of bone, an indirect indicator of bone strength, researchers became increasingly aware of the limitations of such systems in predicting fracture risk.

They turned their attention toward developing a practical, inexpensive, non-invasive way of making strength measurements, a system sensitive enough to monitor and evaluate small changes. The need for such an instrument went way beyond space flight. Since bone deterioration affects a substantial portion of the U.S. population, such an instrument offered broad utility as a research tool for studying bone abnormalities caused by disease, aging and disuse, and as a means of evaluating fracture healing.

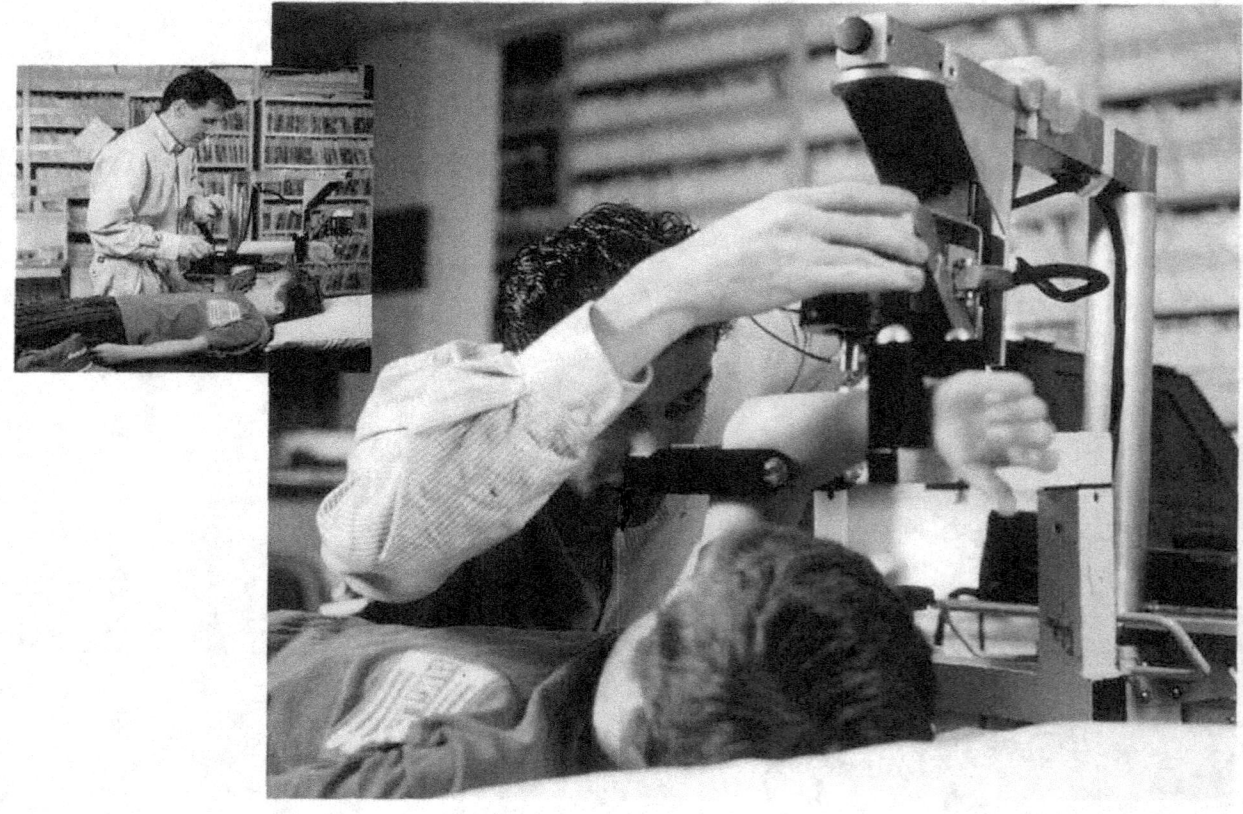

Upper left: At Louisiana's Ochsner Bone Clinic, a technician is measuring the stiffness of a young patient's ulna in a research project aimed at developing advanced methods of treating bone disorders. Shown above is the spinoff Mechanical Response Tissue Analyzer that makes possible, for the first time, direct measurements of bone flexibility.

Now, for the first time, there exists a commercially available instrument to serve those needs. Called the Mechanical Response Tissue Analyzer (MRTA), it is the result of a three-way collaboration among Ames Research Center; Stanford University, Palo Alto, California; and Gait Scan, Inc., Ridgewood, New Jersey, a small business. Beginning in 1977, NASA Ames and Stanford University teamed on a research program aimed at developing impedance devices capable of measuring the mechanical stiffness of human bones. Bending stiffness is a property of the bone that reflects the materials in the bone and its shape; stiffness can be correlated to bone density and calcium content, thereby providing useful information for research on astronaut disuse osteoporosis and related bone disorders among the general population.

After more than a decade of effort, the team came up with a workable device and conducted clinical testing at Stanford University Orthopedic Hospital. In 1989, Ames and Stanford were joined by Gait Scan, which invested its own funds in refining the technology, developing a practical and affordable instrument, and bringing it to the commercial market.

The resulting MRTA is a portable device that detects the bone's response to a brief vibratory stimulus and produces a measurement of the bending stiffness of the ulna (arm) and the tibia (lower leg), the parts of the human body directly involved in weight lifting and weight bearing activities. A technician places a small probe on the skin surface of the limb to be tested and a five-second electrically-induced vibration is applied. The frequencies from the resonating bone are detected at the same site as the stimulator and analyzed by the system's computer, which produces an accurate reading of the bone's bending stiffness.

"The major attraction of the technology," says Dr. Sara Arnaud of Ames Research Center's Life Sciences Division, "is the speed and simplicity with which the measurement gives a complete picture of bone strength." Among other advantages are its safety—MRTA uses no radiation as do some other methods of examining bone—and its cost, about $20,000, which makes it relatively inexpensive as medical systems go.

The MRTA has a wide range of potential applications. It will be used for the original purpose: astronaut postflight monitoring. Dr. Arnaud is using the device to measure tibia strength among working women at Ames. Gait Scan is pursuing applications in monitoring the effects of exercise and rehabilitation on bone stiffness and in osteoporosis, the underlying cause of some 1,300,000 bone fractures each year that involve treatment costs estimated at close to $4 billion.

An application of particular interest is MRTA's use by the Ochsner Bone Clinic, New Orleans, Louisiana, a facility dedicated to research and treatment of patients with osteoporosis and metabolic bone disorders. In a project headed by clinic co-director Dr. Alan Burshell and Dr. Steven Smith, the MRTA is being used in a study of *osteogenesis imperfecta*, or O.I., a disease characterized by brittle bones and increased risk of fracture.

The project involves the cooperation of six generations of a large Louisiana family, the Heberts, who have a history of hereditary O.I. More than 50 members of the family have O.I., although others are not afflicted. One single household has sustained more than 80 fractures; an individual in another family has broken more than 50 bones.

The Ochsner team uses the MRTA to measure bone flexibility, then it compares results among family members with O.I. and those without it. MRTA measurements are further compared with data from CT scans, bone density readings by other instruments, and biochemical/genetic information from blood, urine and skin analysis. Dr. Burshell hopes that this mass of information will lead to advanced treatments for O.I., osteoporosis and other bone disorders.

Programmable Pacemaker

An advanced Trilogy™ cardiac pacemaker that incorporates multiple NASA-developed technologies provides physicians with unprecedented programming capabilities, plus more detailed information on the patient's health and the performance of the pacing system.

Introduced in 1995 by Pacesetter Systems, Inc., Sylmar, California, the Trilogy family of pacing systems represents a fourth generation advancement of the programmable unit first developed in the 1970s by NASA, Johns Hopkins Applied Physics Laboratory and Pacesetter Systems.

In the latter 1970s, Pacesetter brought to the commercial market three significant advances based on the NASA/APL/Pacesetter collaborations: the first rechargeable, long-life pacemaker battery, based on technology for spacecraft electrical power systems; the first single-chip pacemaker, a product of space microminiaturization technology that allowed a substantial reduction in the size of implantable pulse generators; and the first pacing system to utilize bidirectional telemetry, the NASA-developed technology for two-way communication with satellites that provided a way for physicians to communicate with an implanted pacemaker and reprogram it without surgery.

In 1979, Pacesetter introduced the first commercially available bidirectional telemetry pacing system and embarked upon an extensive program of research and development that established the company as a world leader in the very large bradycardia (slow heartbeat) market. Over the years, the company has developed, refined and improved a full line of cardiac pacing products. Pacesetter manufactures all the components of its pacing systems: the pulse generator, or pacemaker, which controls the heartbeat; the pacing leads, which connect the pacemaker to the heart; and the external programmer, which gathers information from the pulse generator that enables assessment of the system's performance and reprogramming.

The new Trilogy family was designed, in consultation with an international board of medical pacing experts, to address concerns that increasingly sophisticated pacemakers were becoming more difficult to program and manage. The Trilogy design features a powerful microprocessor that allows more functions to be fully automatic; this "automatic intelligence," developed by the company's advanced PDx™ diagnostic and programming software, permits a Trilogy unit to adjust many of its settings on the basis of information it gathers about heart function. The net results are enhanced diagnostics and easier programming.

Pacesetter, Inc. is developing a next generation of pacemakers, to be known as Affinity™, in which the number of circuitry components will be halved yet allow incorporation of automaticity, expanded data memory and a sophisticated, handheld PC-based programmer. Originally part of Siemens AG, Pacesetter became—in 1994—a division of St. Jude Medical, St. Paul, Minnesota, a leading manufacturer of medical devices for the cardiovascular market.

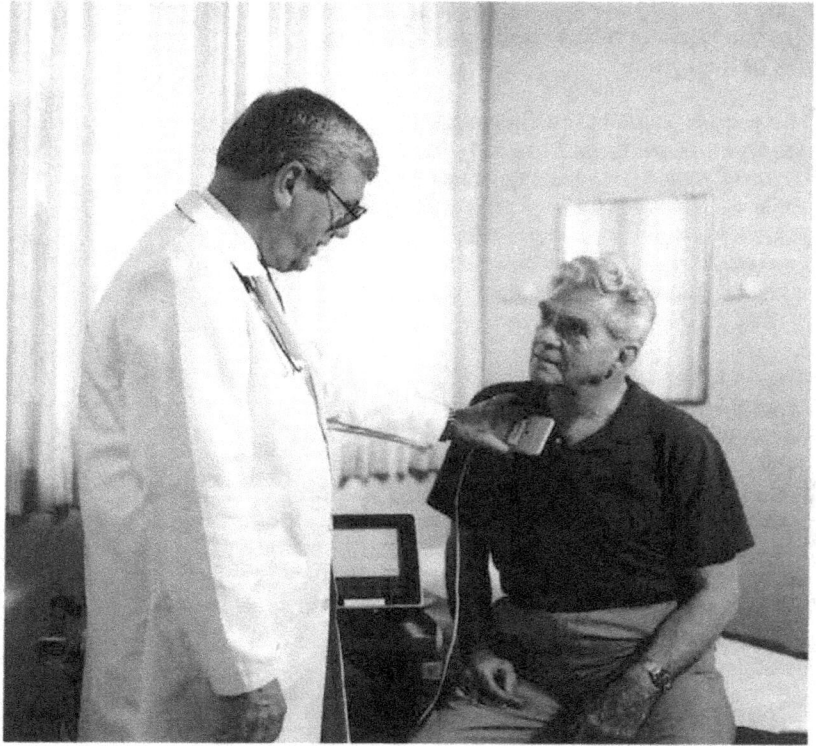

A physician checks a patient's advanced Trilogy pacemaker.

™Trilogy, PDx and Affinity are trademarks of Pacesetter, Inc.

Cardiac Monitor

The basic method of assessing heart function is thermodilution, a procedure that involves insertion of a catheter into the pulmonary artery and is demanding in terms of cost, equipment and skilled personnel time. For monitoring astronauts in flight, NASA needed a system that was non-invasive and considerably less complex.

In 1965, Johnson Space Center contracted with the University of Minnesota to explore the then-known but little-developed concept of impedance cardiography (ICG) as a means of astronaut monitoring. A five-year program led to the development of the Minnesota Imped- ance Cardiograph (MIC), an electronic system for measuring impedance changes across the thorax that would be reflective of cardiac function and blood flow from the heart's left ventricle into the aorta. NASA separately contracted with Space Labs, Inc., Van Nuys, California for construction of space qualified miniaturized impedance units based on the MIC technology. The system was introduced to service aboard Space Shuttle flight STS-8 in 1983.

ICG clearly had broad potential for hospital applications but further development and refinement was needed. A number of research institutions and medical equipment companies launched development of their own ICGs, using the MIC technology as a departure point. Among them were Renaissance Technologies, Inc., Newtown, Pennsylvania and Drexel University of Philadelphia, who jointly devel- oped the IQ System®. The system provides a simple, repeatable, non-invasive way of assess- ing cardiac function at dramatically reduced cost; Renaissance states that the cost of the thermodilution technique runs five to 17 times that of IQ monitoring. The IQ System is in wide use in hospital Intensive Care Units, emergency rooms, operating rooms and laboratories in the U.S. and abroad.

IQ has two basic elements: the non-invasive, disposable patient interface known as IQ-Connect and the touch screen monitor, which calculates and displays cardiac output values and trends. The hardware design of the original MIC was retained but IQ has advanced

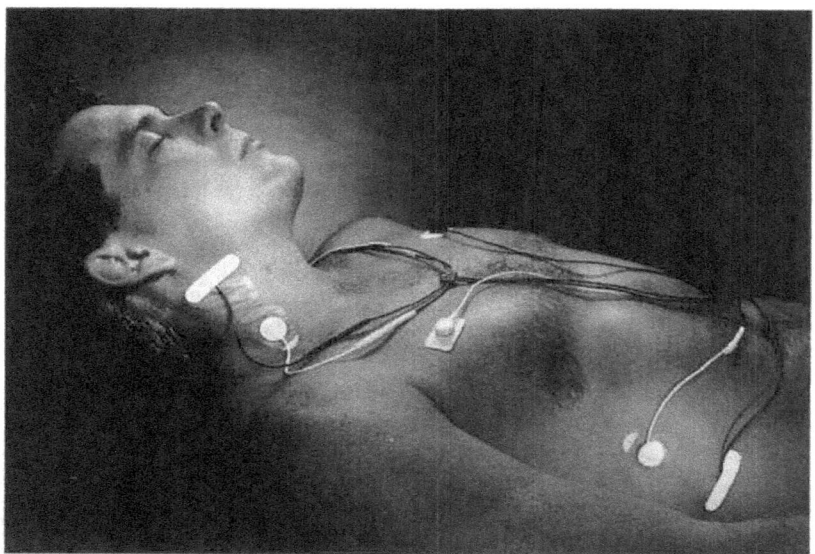

The IQ-Connect interface electronically measures impedance changes across the thorax to reflect heart function.

automated software that features the signal processing technology known as TFD (Time Frequency Distribution). TFD provides three- dimensional distribution of the hemodynamic (blood circulation force) signals being mea- sured, enabling "visualization of the changes in power, frequency and time." This clinically proven capability allows IQ to measure all cardiac events without using estimation tech- niques required in some earlier systems.

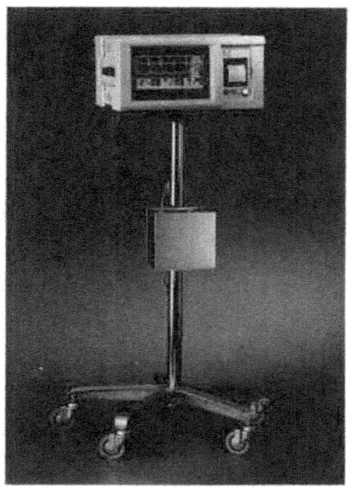

A monitor calculates and displays cardiac output values.

®IQ System is a registered trademark of Renaissance Technologies, Inc.

Telemedicine Program

Telemedicine is the interactive transmission of medical images and data to provide better health care for people in remote or "medically underserved" locations. The concept has been around since the 1920s and its general viability has been demonstrated since the 1950s, but wide adoption was slowed by high costs and technological shortcomings.

Today the technique is burgeoning under the impetus of snowballing advances in computer, videoconferencing and digital imaging technologies that offer potentially reduced costs along with greater efficiency in transmission and display.

Since the 1970s, NASA has been in the forefront of research and demonstration in the field of telemedicine. NASA has an obvious interest because of the potential of telemedicine in care of astronauts operating beyond Earth orbit in the future. But the agency has a broader interest because it has been mandated by Congress to promote the transfer to the private sector of technologies developed in the course of aerospace research, and many of the technologies that make telemedicine possible were originally developed for acquiring visual information from lunar and planetary spacecraft. NASA is actively engaged in developing new technologies applicable to both space and Earth telemedicine and in spurring broader acceptance of telemedicine by conducting demonstrations of the technique's potential in cooperation with local governments and the medical and industrial communities.

A demonstration of particular interest is an ongoing program, started in 1995 and planned as a permanent operation, in which the University of Texas Health Sciences Center at San Antonio (UTHSCSA) is linked with South Texas Hospital, Harlingen, Texas, 250 miles distant. The South Texas area is classified as

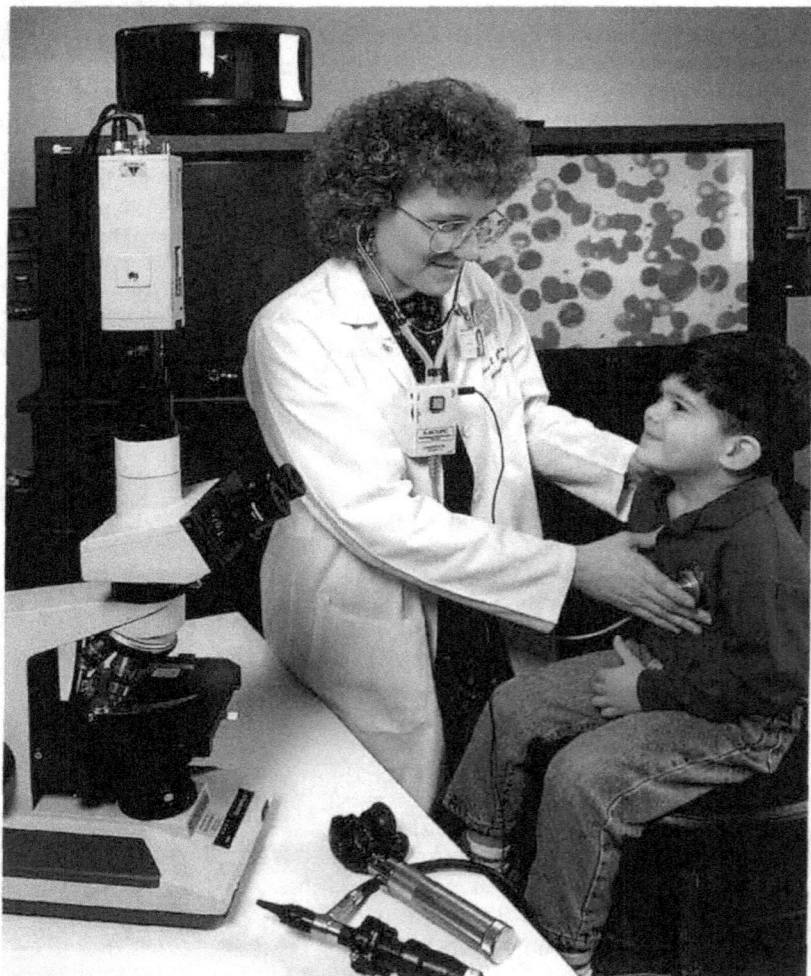

Dr. Terry Lightner transmits the sounds of a child's heart beating to a physician 250 miles away.

From San Antonio, Dr. James Legler views a blood slide in Harlingen, Texas by means of video relay.

medically underserved, with a special problem in the lack of physicians and nurses trained in oncologic (cancerous) diseases. Until the program started, 80 pediatric oncology patients at South Texas Hospital were able to see UTHSCSA cancer specialists only once a month when the specialists visited Harlingen.

Now, via two-way audio/video linkage, UTHSCSA physicians can see and talk with South Texas patients, review laboratory work and consult with doctors providing primary care at South Texas Hospital. They are also conducting 80 hours a month of "teleclinics" and providing instruction in pediatric oncology nursing techniques and family counseling. In addition, South Texas Hospital is getting specialty help in treating the increasing number of tuberculosis cases in the area.

The South Texas project is the result of a broad public/private sector collaboration. In addition to UTHSCSA, South Texas Hospital and the Texas Department of Health, participants include VTEL Corporation of San Antonio, which donated two desktop videoconferencing

systems, maintenance and staff support, and made available two MediaConferencing™ systems at cost; Sprint, Inc., Kansas City, Missouri, which is providing the high speed fiber optic cable link between San Antonio and Harlingen; and Healthcare Open Systems and Trials (HOST) Consortium, Washington, D.C., an industry-based effort to provide rapid application of information technologies to improve health care quality.

NASA contributed expertise to the telemedicine system's design, integration, verification and validation, and additionally provided one-third of the funding to operate the system during its first year. The NASA effort was coordinated by the agency's Office of Space Access and Technology; Dryden Flight Research Center led the effort to integrate off-the-shelf computer and networking systems into the telemedicine link; and Johnson Space Center contributed networking and systems design expertise.

™MediaConferencing is a trademark of VTEL Corporation.

Anti-Shock Garment

Incorporating technology from a NASA project of the 1970s, a new type of anti-shock garment for paramedic use essentially reverses the effect of shock on the body's blood distribution. In shock, the brain, heart and lungs may suffer loss of oxygen because blood accumulates in the lower abdomen and legs; the anti-shock garment applies external counterpressure to the legs and abdomen and returns blood to the vital organs, stabilizing body pressure until the patient reaches a hospital.

Known as DMAST (for Dyna Med Anti-Shock Trousers), the garment was developed by Zoex Corporation, Palo Alto, California and it is distributed exclusively by Dyna Med, Carlsbad, California. There are more than 1,000 suits in use.

Unlike other pressure suits, DMAST is non-inflatable, and it employs lower pressures than other pressure garments for flight and emergency uses. Because it is non-inflatable, it has no tubes, bladders or valves, hence no risk of punctures or leaks. Its simple, one-piece design makes application fast (less than 60 seconds) and easy.

DMAST has demonstrated effectiveness in treating shock from trauma induced by natural disasters or military actions, complications of pregnancy, ruptured internal organs, severe allergic reactions, brain injury and pediatric emergencies. In addition to shunting blood from the patient's legs and abdomen to the heart, lungs and brain, the evenly and sequentially applied counterpressures help curb internal bleeding.

The DMAST technology traces its origin back to 1971, when Ames Research Center was developing a prototype pressure suit designed to protect hemophiliac children from bleeding into elbow and knee joints by straightening and compressing the joint until medical attention was available. Lack of funding forced Ames to curtail the project, but Zoex Corporation picked up the development. Sheri Hillenga, one of the Ames team that had worked on the suit, joined Zoex.

Zoex adopted the low pressure approach and, after years of research, developed a garment that worked successfully. The company obtained patents for the suit in 1992 and the following year teamed with Dyna Med, a medical supply catalog company, to market the product. Sheri Hillenga, operating as VMH Visual Communications, assists Zoex in a marketing capacity and conducts public service demonstrations of the anti-shock garment.

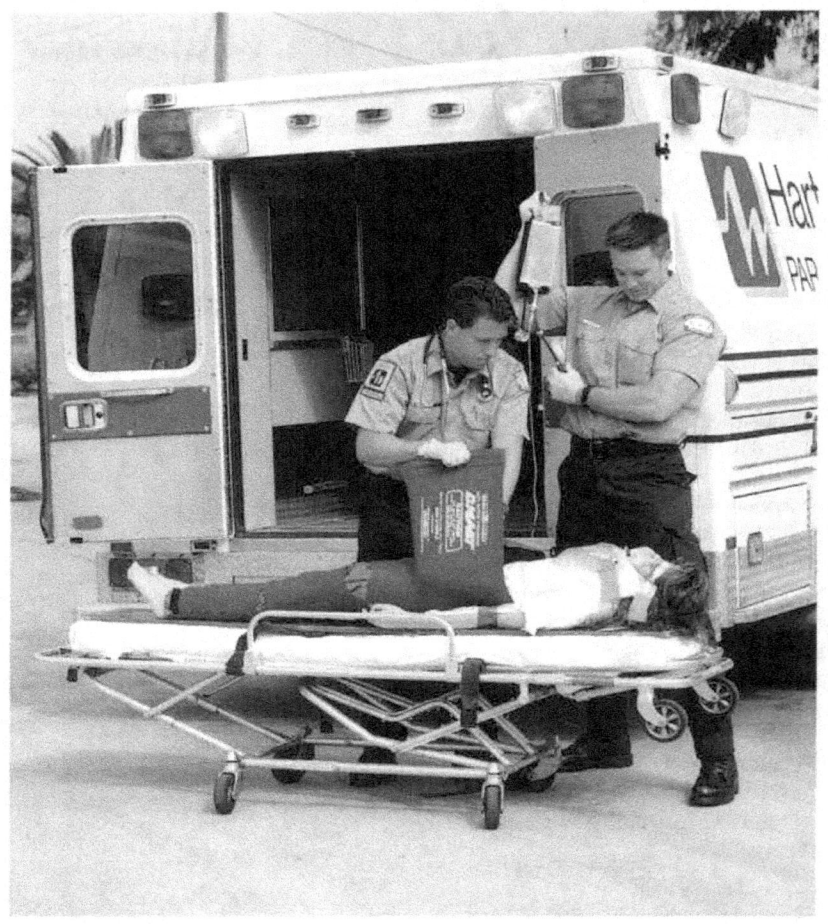

Paramedics fit a shock victim with a Dyna-Med anti-shock garment.

Balance Evaluation Systems

NeuroCom International, Inc., Clackamas, Oregon, a leading manufacturer of medical devices for patients with balance-associated problems, produces the EquiTest® System, a clinical tool which offers utility in diagnosis and in training of patients with balance and mobility disorders.

Both systems are based on core technologies developed under NASA funding. They are "computerized posturography" machines that measure patient responses to movement of a platform on which the subject is standing or sitting, then provide computer-generated assessments of the patient's postural alignment and stability. Widely used by major medical centers in the U.S. and abroad, they have applicability to diagnosis and treatment of such conditions as head injury, stroke, chronic dizziness, heightened risk of falling, and vestibular and central nervous system disorders.

NeuroCom's Balance Master is used by NASA to measure the equilibrium of Space Shuttle astronauts on return from orbit, and Russia recently installed the equipment for use in its space program. In medical testing and rehab use, a patient sits or stands on the platform and works with a special computer training game designed to address specific balance problems. The patient is instructed to move his body in ways that control a small figure on a computer monitor; the goal is to make the figure reach targets on the screen. Therapists can design targets to encourage specific movements by the patient that will help build physical stability, endurance and confidence. Repeated testing provides both patient and therapist an indication of progress.

Now chairman of NeuroCom, Lewis M. Nashner, Sc. D., founded the company in 1984 with initial research grant support from NASA. The original grant helped support development of EquiTest; subsequent funding under the Small Business Innovation Research program supported development of the Balance Master.

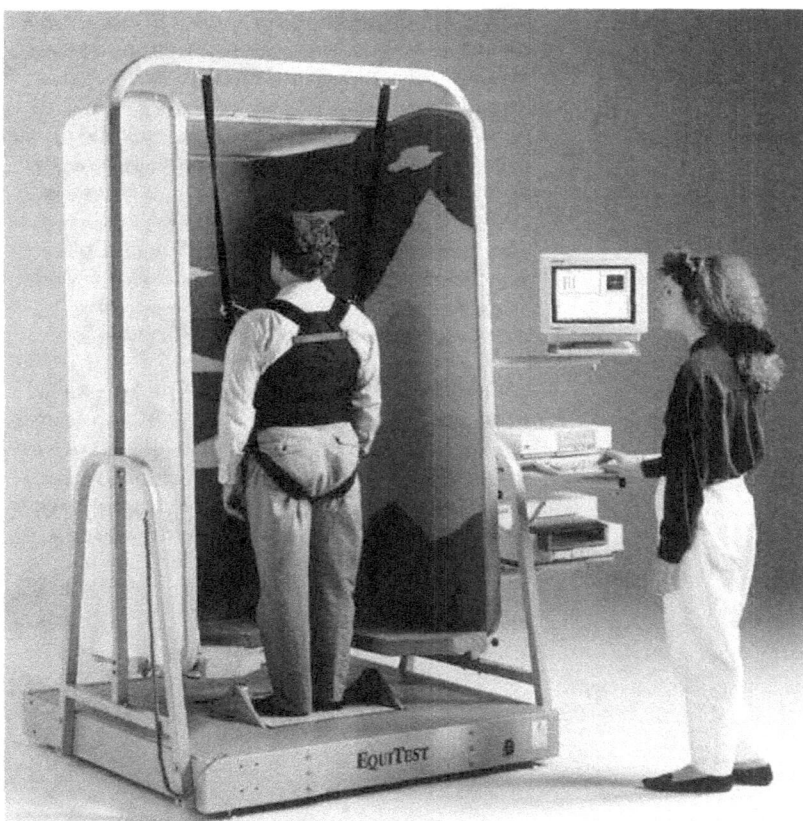

Above: The EquiTest System evaluates balance under dynamic test conditions.

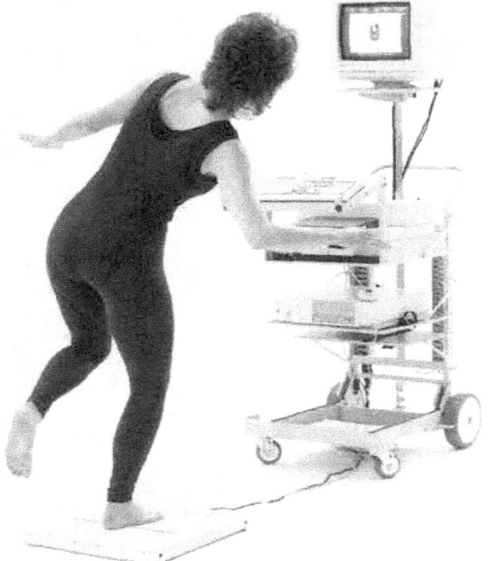

The Balance Master is a tool for diagnosis and training of patients with balance disorders.

A Giant Step in Jetliner Propulsion

A cleaner, quieter, more economical aircraft engine heads technology transfers in the field of transportation

In November 1995, the General Electric GE90 turbofan engine made its flight debut aboard a Boeing 777 jetliner operated by British Airways. Developed and manufactured by GE Aircraft Engines (GEAE), Cincinnati, Ohio, in concert with program participants in France, Italy and Japan, the GE90 is one of the most powerful air breathing engines ever flown. It is also one of the most efficient, one of a trio of advanced technology, very high thrust jetliner engines that offer new levels of operating economy and environmental acceptability.

The GE90's history underlines the lengthy time and large investment—$1.5 billion in this instance—required to bring a major aerospace system from concept to service entry. By the time it was certified in 1995, the engine had been several years in development. Actually, though, some of the key technologies incorporated in the GE90 trace their roots to joint GE/NASA research in the 1970s.

The GE90 was initially certified at 84,700 pounds thrust, but it is capable of thrust levels well beyond that. Only two engines are required to power the big widebody 777, which approaches in size the four-engine Boeing 747 jumbo jet.

It is a type of engine known as a high bypass turbofan. Used in most modern airliners, the turbofan is a propulsion system in which some of the air taken in is compressed, burned in a combustor and expelled to generate power for driving the fan and compressor. A greater amount of the air bypasses the combustion process. In the GE90, the relatively cool bypass air is pushed rearward by a huge (123-inch-diameter) multibladed fan to mix with the hot exhaust gas; the result is a very large gain in overall thrust with minimal fuel expenditure.

A GE Aircraft Engines technician is dwarfed by the mammoth GE90 jetliner engine, which incorporates multiple technologies developed in joint GE/NASA research programs. The large sphere is a test structure.

Propulsion engineers use the term "bypass ratio" to indicate how much of the ingested air bypasses the combustion chamber; generally speaking, the higher the ratio, the more efficient the engine. The GE90's very high 9:1 ratio makes a big contribution to noise and fuel burn reductions. Specifically, the GE90 features a fuel consumption rate 10 percent better than the large commercial turbofans of the pre-1995 generation; a noise level two decibels lower; and emission reductions of oxides of nitrogen (35 percent), carbon monoxide (25 percent) and unburned hydrocarbons (60 percent).

These features have great attraction for airline operators. Fuel is one of the major elements of total operating costs, and fuel efficiency of the order the GE90 offers is vitally important to the world's airlines, who are just beginning to recover financial equilibrium after years of heavy losses. The engine's environmental characteristics provide a valuable bonus, not only in community good will but in a further contribution to improved airline earnings through avoidance of noise and emissions taxes now being levied by foreign nations.

Although GEAE spent years refining them, the basic fuel consumption and environmental improvement technologies that provided the springboard for what eventually became the GE90 stemmed from the company's participation in two NASA research programs.

The first, initiated in 1969 and continuing through the 1970s, was the Quiet Clean Short-haul Experimental Engine (QCSEE) program conducted by Lewis Research Center. QCSEE focused on then-advanced technologies to lower engine noise and address the most troublesome aircraft-emitted contaminants. The program was eminently successful; ground tests of research engines in the 40,000-pound-thrust class demonstrated noise reductions 8-12 decibels (60-75 percent) below the quietest engines in civil transport service. They also demonstrated new technologies to effect sharp reductions in emissions of carbon monoxide and unburned hydrocarbons.

Also in the 1970s, GEAE joined with NASA in a joint Energy Efficient Engine (E³) program managed by Lewis Research Center. Like QCSEE, E³ targeted emission reductions, but emphasized new design techniques for minimizing fuel burn. Highlighting that program was development of a new type of compressor core and an advanced combustor. The GE90's compressor and dual-dome combustor are direct descendants of technology developed in the E³ program and thoroughly proven in extensive ground tests; they are the principal factors in the engine's economical fuel burn, reduced emissions and low maintenance cost features.

In developing the GE90, the company drew upon technology gained from multiple sources, including the NASA experience; expertise acquired in building high thrust systems for military aircraft; and GEAE's development and manufacture—in cooperation with SNECMA of France—of the CF6 and CFM56 families of engines, which have years of service with many of the world's airlines.

The development effort began in the late 1980s and progressed to hardware test in 1992. The ground and flight test program was the most exhaustive ever undertaken by the company, involving some 7,600 hours and 19,000 cycles of endurance testing to simulate more than two years of actual airline experience and maintenance.

Thrust levels of the GE90 series continue to increase. The first growth model—the GE90-92B—achieved its rated thrust of 92,000 pounds in the spring of 1996. Research engines have topped that; the GE90 has operated for more than 150 hours at thrust levels above 100,000 pounds, and has demonstrated a thrust capability of 110,000 pounds. The engine is designed to power all models of the 777 in development or planned, plus other subsonic commercial widebodies contemplated for introduction over the next 20 years.

Advanced Lubricants

Sun Coast Chemicals (SCC) of Daytona Inc., Palm Court, Florida was founded in 1989 by race car driver Edward "Buck" Parker to help alleviate problems on the NASCAR Racing Circuit, such as heat and wear damage to engines and transmissions of race cars. The company's first product, X-1R Concentrate Friction Eliminator, established Sun Coast's reputation as a manufacturer of effective lubricants. In 1994, when Lockheed Martin Space Operations (LMSO) was looking for a more environmentally-friendly lubricant for a mammoth transporter, the company turned to Sun Coast for assistance.

LMSO is NASA's contractor for launch operations at Kennedy Space Center (KSC). The huge transporter is the Mobile Launch Platform, a six million pound "crawler" that moves the complete Space Shuttle from KSC's Vehicle Assembly Building to the launch pad at one mile per hour. The company asked Sun Coast to develop a spray lubricant that would eliminate environmental drawbacks of lubricants then in use yet meet demanding specifications and provide the crawler with extreme lubrication protection. Working in cooperation, LMSO and Sun Coast developed the Crawler Track Lubricant (CTL) in eight months and the formulation passed all NASA tests, met all Environmental Protection Agency requirements, and has since become a major part of the Crawler Preventive Maintenance Program at KSC.

Using the NASA formulation as a departure point, SCC has created and introduced to the commercial market three spinoff products. The first was TTL (Train Track Lubricant), a modified version of CTL that solved a wear problem for the Florida Power Corporation Railroad System. A second spinoff, also in service with Florida Power, is the PSL (Penetrating Spray Lube), which is used in applications where spray lubrication is needed for rust prevention, for loosening corroded bolts, and for lubricating joints and hinges. The third offshoot of CTL is the BHF (Biodegradable Hydraulic Fluid), which has an oxidation life of 10,000 hours.

SCC produces a broad family of environmentally friendly lubricants for applications ranging from industrial assembly lines to oiling a child's bicycle chain. The company has achieved explosive success in a short time and now has distribution networks in 52 countries.

Sun Coast Chemical's Chris Fornili sprays railroad tracks with a special lubricant derived from a lube product developed for NASA's Space Shuttle transporter.

Energy Storage System

Founded in 1985, SatCon Technology Corporation, Cambridge, Massachusetts develops and markets intelligent electromechanical products for aerospace, industrial, transportation and utility applications. SatCon has worked on more than 30 projects with seven NASA centers under the Small Business Innovation Research (SBIR) program. Several of those research assignments, in particular two related to energy storage research sponsored by Lewis Research Center and Marshall Space Flight Center, yielded innovative technology that was later incorporated in SatCon's commercial Flywheel Energy Storage (FES) system.

A flywheel is a chemical-free mechanical battery that harnesses the energy of a rapidly spinning wheel and stores it as electricity with 50 times the storage capacity of a lead-acid battery. Much of SatCon's work for NASA is directed at developing FES systems for spacecraft attitude control and momentum recovery; one development combines energy storage and spacecraft control functions in a single FES system.

In commercial use, FES systems have great potential for adding long life and extended range to electric and hybrid-electric vehicles by providing extra power for acceleration or hill climbing, and by recovering energy normally lost in braking.

In industry and utility applications, FES systems provide continuous electrical power to critical machines and operations during lightning strikes and utility line faults. FES systems also solve quality problems caused by modern, high frequency switching power supplies. Additionally, in utility service, FES systems provide off-

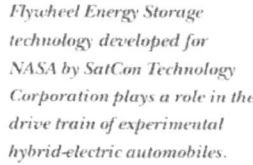

peak storage options for reducing power generation requirements to meet peak power demand.

In August 1995, SatCon signed an agreement with Westinghouse Electric Corporation, Pittsburgh, Pennsylvania for further development and commercialization of SatCon's flywheel technologies. The companies will work together to provide low voltage solutions, including energy storage, for a variety of industrial and commercial users.

Flywheel Energy Storage technology developed for NASA by SatCon Technology Corporation plays a role in the drive train of experimental hybrid-electric automobiles.

The SatCon Flywheel Energy Storage system provides 50 times the energy storage capacity of a conventional lead-acid battery.

Robotic Waterjet Systems

In the late 1970s, before the Space Shuttle made its flight debut, NASA recognized a problem in the fact that the Shuttle's two solid rocket boosters were reusable; they were designed to operate for two minutes on a Shuttle mission, then separate and drop Earthward to a parachute-softened splash in the ocean for recovery and refurbishment. The problem lay in the refurbishment process: how to strip away multiple layers of tough, removal-resistant paint and thermal protection material, and do it in such a manner that the booster's casing would not be damaged. Clearly, NASA saw, something substantially more advanced than conventional manual stripping would be required.

Marshall Space Flight Center (MSFC) teamed with United Technologies' USBI Company, Huntsville, Alabama, NASA's contractor for booster refurbishment, to develop an advanced stripping system based on hydroblasting, or high pressure waterjet cleaning. To compress eight years of complex research and innovative development into a paragraph, the NASA/USBI team developed a waterjet capable of slicing through the thermal protection material and blowing away the particles; found a way to regulate the jet stream to strip paint and primer beneath the outer protective coating, one layer at a time; and developed a computer-directed six degrees of freedom robot for precise control of the waterjet, in order to prevent damage to the casing. The end result was a system that could strip a booster in a fraction of the time required by manual methods.

This technological solution and its subsequent commercialization is a prime example of high value spinoff from aerospace technology (in this case NASA and, later, Department of Defense technology). It is an example of a technology transfer that not only resulted in formation of a new company, but also spawned a whole new industry. The innovative technology offers a broad range of benefits to maintenance and overhaul organizations in cost and time savings, worker protection and a variety of environmental protection advantages.

United Technologies Corporation took advantage of the apparent commercial potential of the system by investing company funds in refining the NASA/USBI technology, advancing it in new directions, and—in January

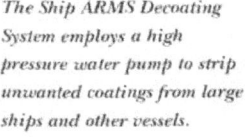

The Ship ARMS Decoating System employs a high pressure water pump to strip unwanted coatings from large ships and other vessels.

1993—establishing a company to commercialize and market the technology. The company is Waterjet Systems, Inc., Huntsville, Alabama, operated by United Technologies' Pratt & Whitney division. Waterjet Systems pursued the military potential of the technology under an Air Force MANTECH (manufacturing technology transfer) program to adapt the booster-stripping technology to large aircraft, then extended the technology to ships through a Navy MANTECH program.

With these and other extensions of the technology, Waterjet Systems established a family of hydroblast systems under the general designation ARMS® (Automated Robotic Maintenance Systems). Generally, they employ high pressure water pumps to create waterblast streams at 55,000 pounds per square inch that are controlled by target sensitive robots maneuvering precisely over one-of-a-kind surfaces. The ARMS family includes:

Aircraft ARMS®. The first application of Aircraft ARMS, a fully automated version derived from the NASA/USBI technology and refined under the Air Force MANTECH program, is the LARPS (Large Aircraft Robotic Paint Stripping) system for Air Force use. Introduced to operational service in 1995, its goals are to reduce stripping time by 50 percent and use of hazardous chemicals by 90 percent. Aircraft ARMS systems are designed for painting and depainting commercial as well as military aircraft.

Ship ARMS®. The Ship ARMS Decoating System was designed for use on large ships, boats, barges, floating drydocks and other vessels. It is a semiautomated, totally mobile system that includes a trailer-based high pressure high flow water pump, a precision computer-designed nozzle and a specialized robotic manipulator and end effector. The system not only strips a ship's multilayer anticorrosion and antifoulant coatings, one layer at a time, but also captures all process effluent; a specially designed vacuum shroud and water recovery/reuse system collects stripped materials and prevents hazardous materials from running off into the water or onto the dock. The U.S. Navy was the first

customer; the process has been demonstrated on the aircraft carrier *USS Nimitz* and two other ships. Among a lengthy list of benefits are environmental gains, such as no airborne particles, reduced waste disposal, reduced eco-impact, and cost-related gains, such as no need for scaffolding or masking, shorter strip and drydock time, and less cleanup.

Engine ARMS®. Turbine engine parts—vanes, combustors, blades, for example—have a variety of coatings to protect them from the hot operating environment. At overhaul time, the engine is disassembled and the parts are stripped of their coatings, customarily by acid bath and grit blasting, processes that are both labor intensive and hazardous to personnel. Engine ARMS does the job faster and cheaper with bonuses in reduction of toxic chemicals, waste disposal and human protection equipment.

Component ARMS®. Components in this sense means major segments of aircraft or military/industrial systems. Among examples that can be decoated by Component ARMS are engine cowlings, rudders, flaps, radomes, helicopter blades, horizontal stabilizers, tail cones and small vehicles (Humvees, armored personnel carriers, tanks, etc.).

°ARMS, Aircraft ARMS, Engine ARMS, Ship ARMS and Component ARMS are registered trademarks of United Technologies Corporation

The decoating system can reach high surfaces, eliminating the need for scaffolding.

Aircraft Ducting

As part of their technology transfer programs, NASA centers frequently provide assistance to industrial firms who request it, for example, analysis of a new product design, help in solving manufacturing problems, or product testing in specialized types of facilities that are not available commercially. In many cases, NASA's work results in improvements to the product.

An example of a company that benefited from such assistance is Templeman Industries' TI Aerospace Systems, Inc., North Chicago, Illinois, a manufacturer of composite ducting for aircraft.

TI had developed a composite environmental air duct, known as the Ultra-Seal Ducting System™, that offers a 50 percent weight reduction in comparison with existing metallic ducting. The ducting is intended primarily for commercial aircraft, but it also has utility in a number of military applications.

TI had to assure that the air duct design would meet environmental and cyclic stress specifications, but the company was unable to find a commercial firm with adequate test capabilities. TI turned to Marshall Space Flight Center (MSFC) and the center's Component Test Branch of the Propulsion Laboratory agreed to conduct a structural evaluation.

MSFC put the duct system through tens of thousands of cycles of pressure and leakage tests that were collectively equivalent to 86 years of aircraft takeoffs and landings; the ducting performed satisfactorily. The system is now in production and in use by aircraft manufacturers, among them the two largest U.S. commercial jetliner builders, Boeing Commercial Airplane Group and McDonnell Douglas Corporation.

™Ultra-Seal Ducting System is a trademark of Templeman Industries.

Templeman Industries' J.B. Templeman (left) and Scott Templeman conduct a quality control test on an Ultra-Seal Duct component.

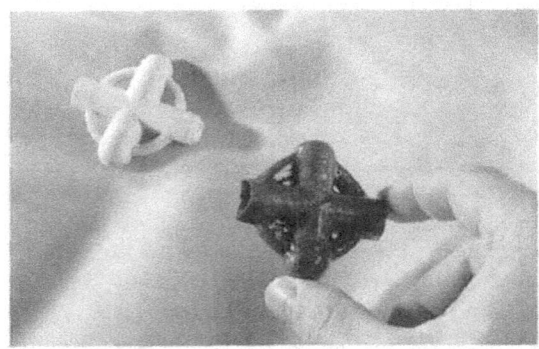

A proprietary casting technique allows fabrication of small and intricate yet high performing Ultra-Seal Duct parts.

Night Vision Camera

Low light-level viewing devices are typically used by the military services for surveillance and intelligence-gathering activities at night or in conditions of poor visibility; they are additionally used in such applications as medical imaging and spectroscopy. Conventional low light TV imaging systems generally employ image intensifiers together with Charged-Coupled Devices (CCDs), cameras that capture scenes electronically (without film) and produce a digital image with relatively high resolution.

PixelVision, Inc., Beaverton, Oregon has introduced a new Night Video™ NV652 Back-illuminated CCD Camera that operates without an image intensifier, thereby freeing the system of certain limitations imposed by the intensifier, yet it is capable of acquiring quality images at low light levels previously attainable only with image intensifier tubes; the development drew upon technology developed by Scientific Imaging Technologies, Inc. (SITe), also of Beaverton, and on the expertise of a longtime CCD developer.

Conventional video cameras use front-illuminated CCDs that impose some limitations on performance. The Night Video NV652 system illuminates and collects charge through the back surface; this design permits the image's photon to enter the CCD unobstructed, allowing for high efficiency light detection in the visible and ultraviolet wavelengths. In a typical airborne military observation application, the NV652 camera offers advantages over standard intensified CCD sensors, according to scientists who developed the back-illuminated sensor. They include greater resolution under low light conditions through increased sensitivity; better target identification through superior contrast and resolution; lower cost; and longer lifetime through increased reliability.

The back illumination technology that is key to the NV652's sensitivity was developed by SITe; George M. Williams, who worked on the program as a SITe employee, has joined PixelVision as vice president and general manager of the Commercial Systems Division. The NASA technology input was provided by James R. Janesick, PixelVision vice president,

chief scientific officer and director of the company's Advanced Sensors Division; Janesick was formerly with Jet Propulsion Laboratory, where he acquired 23 years of experience in CCD technology and systems design.

The NV652 night vision camera is representative of a broad line of PixelVision back illuminated low light level imaging devices for government, medical, scientific and industrial applications. Vice president Janesick states that the company's focus is on marketing advanced CCD technology for ultra-large, ultra-high speed arrays used in medical, scientific and movie digital camera systems.

The sensor in PixelVision's NV652 Night Video low light level camera makes night flying safer by increasing visibility.

Night Video is a trademark of PixelVision, Inc

Intelligent Highway System

In July 1995, Secretary of Transportation Federico Pena formally opened the control center of the Transportation Guidance System (TransGuide) at San Antonio, Texas, the 10th largest city in the U.S. The ceremony marking the initial operation of the first and most advanced Intelligent Transportation System (ITS) in the U.S. was attended by officials of the Texas Department of Transportation (TxDOT), the Federal Highway Administra-

tion, the Texas state government and the City of San Antonio.

Also in attendance were representatives of some 60 companies that teamed to support the project theme of "Technology in Motion," which refers to the creation of "smart" roadways through application of a number of technological innovations, many of them derived from NASA technology. The contractor team was headed by AlliedSignal Technical Services Corporation, Columbia, Maryland, a leading NASA contractor since the earliest days of space flight.

The TransGuide project began in 1990 when TxDOT began an assessment of advanced technologies applicable to traffic management. The major conclusion of the study effort was that adding lanes to San Antonio's highways was not the answer to efficient expansion; rather, the road network had to be made intelligent by the application of advanced video, sensor, computer and communications technologies. After a competition among several major high tech firms, TxDOT selected AlliedSignal as prime contractor and overseer of the 60-plus specialized companies.

From a control center, workers monitor and manage the road system in San Antonio.

The traffic system operator selects the cameras monitoring a particular area.

AlliedSignal's assignment called for completing the design; procuring, installing and integrating the equipment; testing the system; and training the personnel to manage traffic on what would eventually be a 191-mile freeway network. The initial 26 miles became operational in 1995.

The comprehensive deployment of advanced technologies provides San Antonio with a safer, more efficient highway system, enables expedited traffic flow, cuts auto fuel consumption and emissions, and allows quicker detection of the response to "incidents" (a traffic engineering term for accidents). The system includes pairs of inductive-loop vehicle detectors in each highway lane at half-mile intervals; full motion, full color video cameras at one-mile intervals; 50 fiber optic changeable message signs; 358 overhead lane control signals; and computerized control of traffic signals.

At the core of TransGuide is a command and control system similar to one of NASA's major control centers. Inside the 48,000-square-foot center, video and computer images are displayed on console monitors and on a 60-foot video wall. Each operator also has four video monitors on which to display images from the roadside cameras.

The system monitors the passage of traffic over the sensors embedded in the roadways and—based on traffic speed and flow rates—detects likely incidents. Control center operators are alerted to the occurrence of an accident and the area of the occurrence is highlighted on the map display.

Simultaneously, the system activates video cameras on the roadway to enable the operator to investigate the accident. The operator answers three questions posed by the computer and the system returns a recommended response to minimize the traffic impact of the accident. The system is designed to detect any accident within two minutes (as compared with 20 minutes by manual detection) and to prepare and implement a response in as little as 15 seconds.

The TransGuide ITS incorporates multiple NASA technologies stemming from AlliedSignal's broad experience as a NASA contractor. AlliedSignal Technical Services is NASA's primary services contractor at Goddard Space Flight Center, a subcontractor for mission operations and support to Rockwell International at Johnson Space Center, and a contractor to Jet Propulsion Laboratory. The company also provides services to Ames and

Mounted on poles, the cameras can be manipulated remotely.

Lewis Research Centers and three stations of the Tracking and Data Relay Satellite System.

Allied Signal continues to support the TransGuide software as the system expands and to perform complementary transportation management functions.

A Leak Monitor for Industry

Leading a sampling of spinoffs that enhance public safety is an automated gas leak detection system for industrial applications

In space flight operations, hydrogen leaks pose significant operational and safety problems. In systems that use hydrogen propellants, such as NASA's Space Shuttle, the location and severity of a leak must be determined in a timely manner; failure to do so may breed consequences that range from program delays to loss of life and equipment.

In 1990, leaks on the launch pad necessitated grounding of the Space Shuttle fleet until the leak sources could be identified, a process that was costly both in monetary terms and in disruption of Shuttle schedules. The incident triggered an intensive government/industry/academic effort to develop a comprehensive leak monitoring system for space launch vehicles. And, since industry experienced similar leak detection difficulties, the effort was expanded to a double-barreled approach that additionally embraced development of a commercial automated gas leak system for industrial applications.

The latter system—the Model HG2000 Automated Gas Leak Detection System manufactured by GenCorp Aerojet Industrial Products—made its debut in 1995 at the St. Thomas, Ontario assembly plant of Ford Motor Company, where Ford is producing autos that operate on natural gas in response to the air quality requirements of the 1990 Clean Air Act amendments.

Other objectives of the program, including a system for improving NASA's leak detection capabilities for operation of current and future spacelaunch vehicles, were similarly accomplished. The development of advanced leak sensors and the automated monitoring system won a 1995 R&D 100 Award, presented annually to the 100 most significant technological advances of the year.

At the time the development effort was launched, no commercial system existed for detecting hydrogen leaks at low concentrations in inert (oxygen free) environments with high sensitivity. There were hydrogen sensors available, but they had serious deficiencies when used in applications that demanded operation in inert environments. To address

A natural gas-powered Ford Crown Victoria is being checked for hydrogen leaks.

both matters simultaneously—the sensor and the complete system—the development program progressed along two parallel fronts.

NASA's Lewis Research Center, in cooperation with Case Western Reserve University (CWRU), Cleveland, Ohio, undertook development of advanced point contact hydrogen sensors. An evaluation of sensor technology led to a decision to use a palladium-silver solid state sensor that did not require an oxygen atmosphere to detect hydrogen leaks and could identify concentrations as low as one part in a million. The sensor was to be fabricated on a silicon wafer in a structure that would allow high sensitivity to changes in hydrogen concentrations, and it had to be miniaturized, or "microfabricated," to minimize power consumption and allow placement of sensors in a wide variety of locations. Although the concept of such a sensor had been investigated, it had never been fully developed as a complete microfabricated package.

The other part of the two-pronged development effort was conducted by Marshall Space Flight Center and GenCorp Aerojet Industrial Products, Las Vegas, Nevada. Their job was to develop a complete microprocessor-based hardware/software system to monitor multiple microfabricated hydrogen sensors and visually display the source and magnitude of hydrogen leaks in real time.

Success on both fronts led to accomplishment, in 1994-95, of multiple objectives: a significant advancement of sensor technology; a system for NASA use in leak-checking the Space Shuttle propulsion system; a prototype Aerojet/Marshall system for leak detection in Reusable Launch Vehicles now being developed; and the first commercial application of the system, used in Ford Motor Company production of the 1996 natural gas-powered Crown Victoria auto.

In development tests of the vehicle, which is fueled by compressed natural gas stored at 3,000 pounds per square inch pressure, Ford employed conventional pressure decay and soap bubble techniques that can take hours to determine that leakage is below the acceptable limit (2.5 cubic centimeters per hour). That clearly wouldn't do for production line leak checking. Ford wanted a system that could test all critical joints and provide a leak reading in less than 15 minutes. Additionally, Ford required that its production line system work with nonoxygen-containing gas; make leak measurements 10 times lower than the acceptable rate; provide simultaneous detection at eight locations; and be able to measure leaks during system pressurization. Tests of the Aerojet system showed that it was the only system capable of meeting Ford's specifications.

The NASA/CWRU/Aerojet team is continuing to work on improving the system. CWRU and Lewis Research Center are focusing on increasing the detectable hydrogen concentration range of the sensor in order to broaden the range of useful applications. Aerojet and Marshall are similarly working to improve the complete system and Aerojet is exploring possibilities for additional commercial applications, among them safety monitoring of hydrogen facilities, such as chemical plants and refineries, and monitoring of hydrogen buildup in nuclear waste depositories.

A NASA/industry team developed a hydrogen sensor that fits on a wafer chip.

Nondestructive Test Probe

Under NASA's Aircraft Structural Integrity program, which addresses ways to enhance the safety of aging commercial jetliners by improving structural inspection and lifetime prediction techniques, Langley Research Center is developing a variety of devices and procedures for detecting cracks, disbonds and corrosion during routine aircraft service inspections. One such device is the CrackFinder, now commercially available; it is manufactured by Krautkramer Branson (KB Instruments), Lewistown, Pennsylvania under an exclusive NASA license.

The CrackFinder is an electromagnetic probe for nondestructive evaluation (NDE), developed to meet a need for a highly reliable, low cost method of rapidly scanning aircraft skins for surface breaks. It is based on eddy current technology that enables extreme sensitivity to fatigue cracks in aluminum alloy plates. An eddy current is an electrical current induced by an alternating magnetic field; eddy current techniques are generally used for detecting surface cracks, where ultrasonic systems are used for detecting internal flaws.

The device employs an innovative self-nulling feature that makes the CrackFinder simple yet highly accurate, reliable and economical. Self-nulling means that the device automatically recalibrates to zero so that each flaw detected produces a reading. When the probe is placed on a flaw-free metallic object, its output is automatically nulled. The presence of a flaw alters the delicate self-nulling condition, causing a distinctive probe output amplitude. A bargraph display indicates crack severity.

A subsidiary of Emerson Electric Company, Krautkramer Branson says the innovative probe removes some major limitations of conventional eddy current testing: the CrackFinder is more affordable, costing about one-third the price of a conventional eddy current instrument package; it is small and lightweight, weighing only nine ounces where conventional eddy current instruments weigh six to 20 pounds; its small size and simplicity allow installation of multiple probes in inaccessible locations for periodic monitoring of crack growth in critical infrastructures; and it requires minimal operator training. It also needs no calibration or balance circuitry.

The versatile CrackFinder has applications beyond the aging aircraft market, for example, steel structures, ski lifts and other structures where detection of fatigue is critical. NASA's Technology Application Team at Research Triangle Institute played a part in bringing the CrackFinder to market by identifying Krautkramer Branson as a prospective NASA commercialization partner.

Developed to scan aircraft skins for surface breaks, the CrackFinder is available for industrial use in fatigue detection.

Stress Measurement System

As part of NASA's Aircraft Structural Integrity program (see page 68), Langley Research Center teamed with Stress Photonics Inc., Madison, Wisconsin to develop an infrared-based stress measurement system for use in nondestructive evaluation of materials and structures. Stress Photonics has commercialized the technology and incorporated it into the DeltaTherm 1000™ system.

Both the NASA system, intended for inspection of aircraft structures to detect cracks and disbonds, and the commercial unit make use of the fact that all materials change temperature when compressed or expanded. In solid materials, this effect is known as thermoelasticity. Thermelectric Stress Analysis (TSA) employs special infrared detectors and signal processing equipment to image temperature changes that correspond to the dynamic stresses in a structure; modern TSA systems can measure temperature changes as small as one-thousandth of a degree Centigrade.

Stress Photonics' DeltaTherm 1000 is an infrared differential thermography system for TSA and thermal nondestructive evaluation. The system is used to verify mathematical models of a design, demonstrate the effects of loadings on the design, measure stress concentrations and stress intensity factors in load-bearing components, and survey a structure for potential problem sites. It combines digital signal processing technology with a special infrared camera to provide instantaneous thermal images and live TSA (differential) images. Data collected from the infrared camera is processed at 434 frames a second; processed images are sent to a computer and/or to a video monitor for immediate display.

™DeltaTherm 1000 is a trademark of Stress Photonics Inc

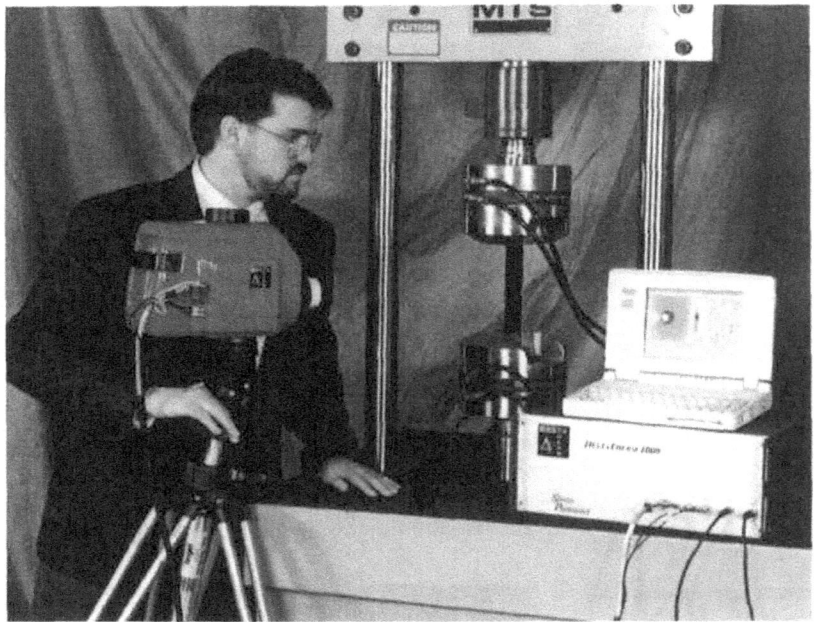

The DeltaTherm 1000 stress measurement system combines digital signal processing technology with an infrared camera.

Used in nondestructive evaluation of materials and structures, the system images temperature changes that correspond to dynamic stresses in a structure.

Voltage Sensor

To demonstrate the feasibility of measuring high voltages employing optical techniques, Lewis Research Center awarded a Small Business Innovation Research contract (see page 108) to SRICO, Inc, Powell, Ohio. The contract called for development of a fiber optic voltage sensor for use in management of aircraft and spacecraft electrical power systems; because it uses glass and light to sense and transmit electricity, fiber optic measurement offers potential for safety and accuracy improvement in voltage measurement.

SRICO successfully developed a NASA prototype device and, using private capital, developed a commercially marketable sensor for terrestrial applications. Among the many commercial uses SRICO cites measurement of electric field and voltage in electric power systems and hazardous environments; lightning detection in avionics and mining; fiber optic communications systems; non-contact probing of high-speed integrated circuits; biomedical engineering and instrumentation; and charge measurement in photocopiers and ion neutralization systems.

The innovative integrated optic voltage sensor employs reverse poling technology that permits the use of simple electrode structures for high voltage sensing without the need for voltage division. The design eliminates electrical isolation problems between the high voltage system and the control system. The sensor and the optical fibers are immune to electromagnetic interference, thus yield accurate measurements over a wide dynamic range.

SRICO's fiber optic voltage sensor offers improved accuracy in voltage measurements.

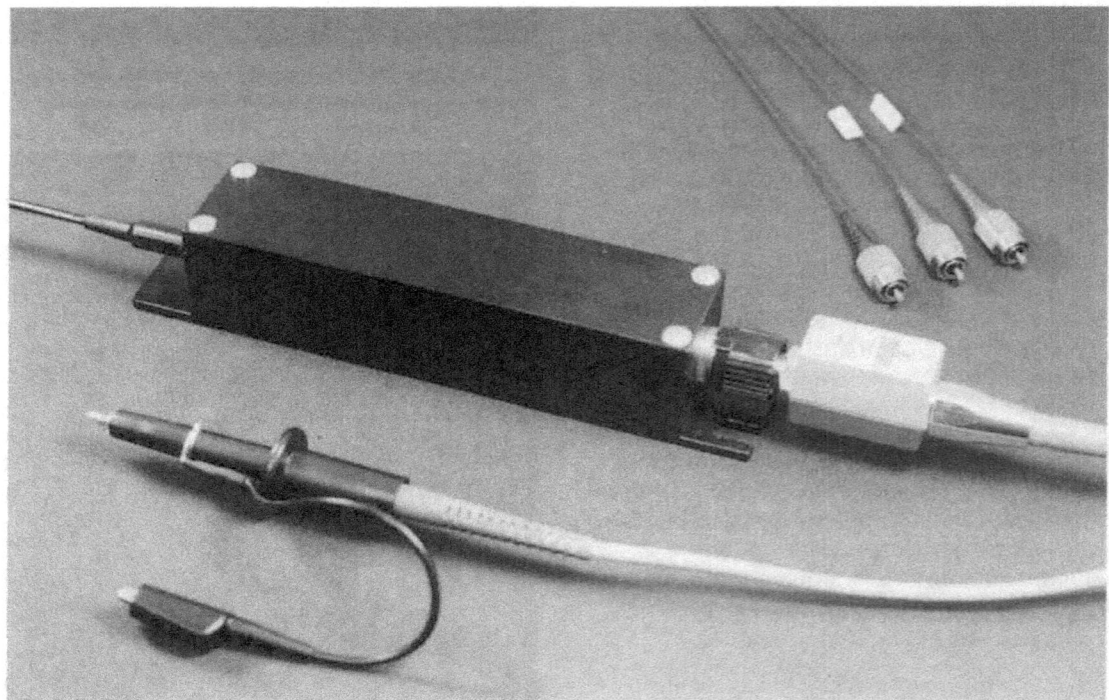

Toxicity Warning Badge

GMD Systems of Bacharach, Inc., Pittsburgh. Pennsylvania manufactures and markets toxic gas detection and air sampling devices, with special emphasis on passive "badge" type dosimeters, worn by workers in potentially toxic environments to provide warning of exposure to hazardous chemicals.

For monitoring NASA and contractor personnel exposed to hydrazine and monomethyl hydrazine (MMH), NASA contracted with GMD Systems for development of a colorimetric gas monitoring dosimeter. Hydrazine and MMH are hyperbolic fuels that ignite on contact with an oxidizer; typical risk areas where exposure monitoring is required include facilities where the fuels are produced and facilities where workers are engaged in assembly and operational handling of spacecraft, missiles and aircraft auxiliary power units in which the fuels are used.

GMD Systems developed a reliable monitoring dosimeter for NASA, then modified the NASA technology to create a new commercial product, the GMD 530 Series Hydrazine Badge. The commercial version is used in chemical manufacturing, industrial cleaning applications and in areas where hydrazine is used as an oxygen scavenger, such as boiler feed water in the electric industry.

The 530 series badge has two separate paper tape chemistries in one badge; a pair of circular windows in the badge allow each tape to be exposed and observed. In the presence of either hydrazine or MMH, both tapes change colors, providing an immediate visual alarm.

In both cases, the density of the color stain increases in proportion of the hydrazine/MMH in the air and the total time of exposure. The system includes two types of pocket-size estimator cards (Dose Estimator Units), one for each substance. A supervisor uses the estimator to match the color of an exposed badge with a set of color standards for the particular substance involved; the standards are shown on the estimator card for quick reference. The color matching serves as a basis for immediate, accurate estimation of the level of exposure.

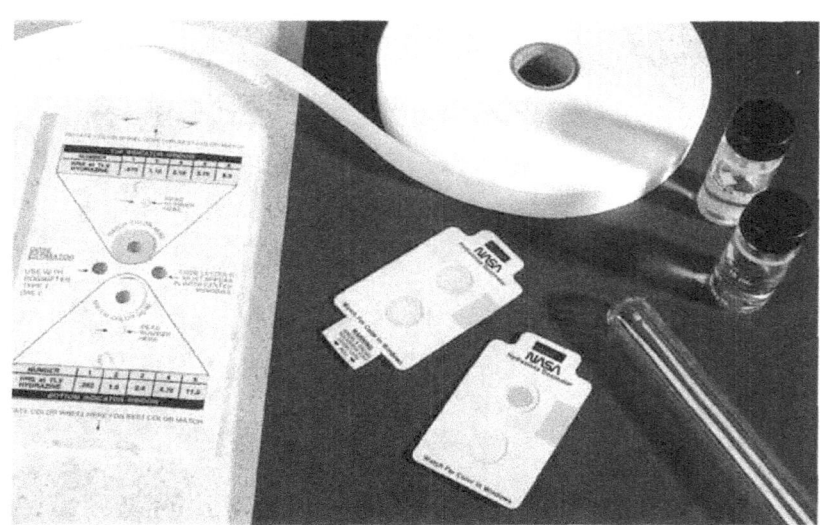

GMD Systems' Hydrazine Badge changes color when its wearer is exposed to a chemical containing a hydrazine component.

The badge offers secondary utility as a low-cost leak detector hung on suspect pipes or valves.

Nutritional Products from Space Research

An algae-based oil, a key ingredient of infant formulas, heads a selection of spinoffs for consumer, home and recreational use

It is difficult to imagine a connection between planetary space flight and baby food, but the connection does in fact exist. Commercially available infant formulas now contain a nutritional enrichment ingredient that traces its existence to NASA-sponsored research that explored the potential of algae as a recycling agent for long duration space travel.

The ingredient is an algae-based, vegetable-like oil known as Formulaid®. It was developed and is manufactured by Martek Biosciences Corporation, Columbia, Maryland, which has pioneered the commercial development of products based on microalgae; the company's founders and principal scientists acquired their expertise in this area while working on the NASA program.

Formulaid is Martek's leading product. Its value as a dietary supplement lies in the fact that it contains two essential polyunsaturated fatty acids known as DHA (docosahexaenoic acid) and ARA (arachidonic acid). DHA and ARA, found in human milk but not in most infant formulas, are believed by many researchers to be associated with mental and visual development.

Martek literature amplifies that statement:

"DHA and ARA are the predominant fatty acids in the grey matter of the brain and DHA is particularly enriched in the retina. Children and adults obtain DHA primarily from their diets, since humans synthesize only small amounts of DHA. Infants acquire DHA and ARA initially *in utero* during pregnancy, and then from their diets via their mothers' milk. DHA and ARA dietary supplementation may be particularly important for premature and low birth weight infants, who may not get their full allotment *in utero*."

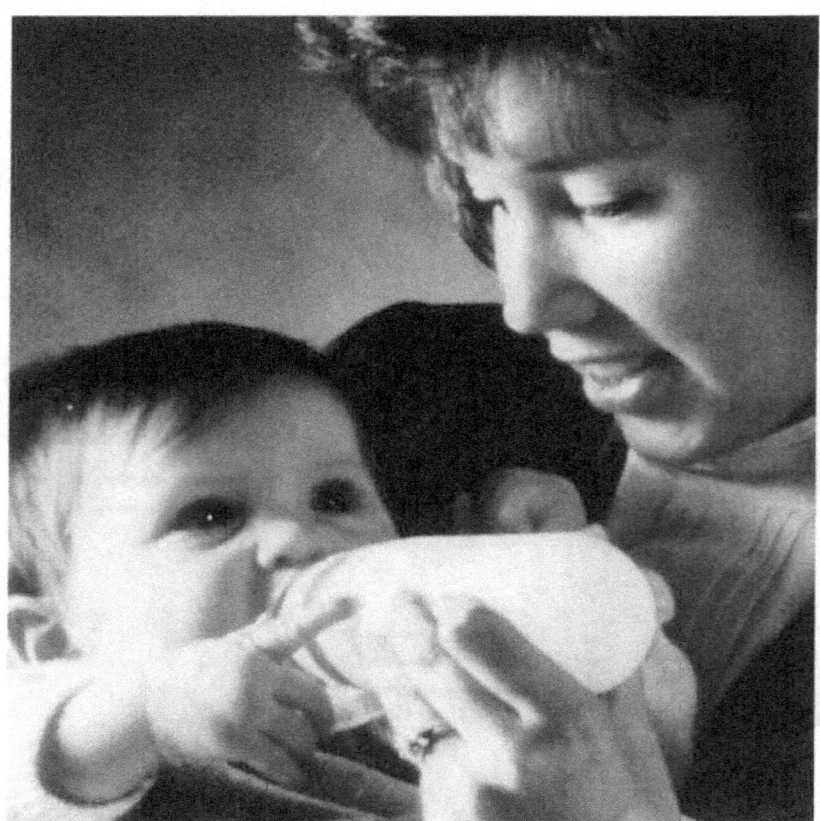

Left: The formula being fed to the baby contains an algae-based additive highly enriched in nutrients believed to be beneficial to infant mental and visual development; called Formulaid, the additive is an offshoot of space research. The additive has been introduced in Europe in the formula Nenatal (right) produced by Nutricia of The Netherlands.

There are also reasons to believe, Martek says, that DHA-supplemented formula can be beneficial to full term infants as well as preterms. Two major nutritional authorities—the British Nutrition Foundation and the United Nations Food and Agriculture Organization/World Health Organization Expert Committee on Human Nutrition—support that belief; both have endorsed use of DHA and ARA as additives in both pre-term and full term infant formulas and as nutrients for women during pregnancy and lactation.

Formulaid and other Martek products had their origins in a NASA program of the early 1980s known as CELSS (Closed Environment Life Support System). Under NASA contract, Martin Marietta Laboratories, Inc., Baltimore, Maryland experimented with the use of microalgae as a food supply, a source of oxygen and a catalyst for waste disposal on future human-crewed planetary missions.

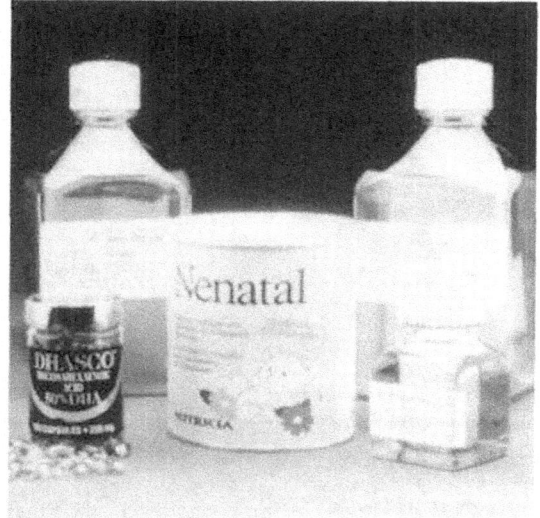

In 1985, Martin Marietta dropped out of the CELSS program when the company decided to discontinue support for bioscience research. Several of the Laboratories' scientists saw an opportunity to start a company and seek to commercialize their expertise in deriving compounds from algae and plants. In May 1985 they left Martin Marietta to form Algatex, Inc., which became Martek Biosciences Corporation in September of that year. Three of the group are still with the company: Dr. Richard J. Radmer, president and chief scientific officer; Dr. David J. Kyle, vice president, head of research and development; and Dr. Paul W. Behrens, director of physiology. Henry Linsert, Jr., is chairman and chief executive officer of Martek.

Martek has devoted more than a decade to research on microalgae, which the company calls "a biochemically rich and diverse kingdom of organisms that have been overlooked by science and commercial interests because of culturing difficulties." Now staffed by some 60 scientists representing all of the life sciences, Martek has solved many of the culturing problems and developed a number of patented products and processes, of which Formulaid is the "flagship" product.

In 1994, Martek was granted a U.S. patent for Formulaid and in the same year the company signed an agreement with Nutricia, a leading European nutritional product manufacturer, whereby Formulaid was to be added to Nutricia's infant formula for sale in Europe. Nutricia successfully introduced the formula for pre-term infants and is in process of expanding Formulaid use to formula for all infants. Martek has also concluded licensing agreements with the Swiss company Sandoz Nutrition Ltd., the Israeli firm Maaborot Products Ltd., and two American infant formula producers, American Home Products and the Mead Johnson Division of Bristol-Myers Squibb. The U.S. licensees are awaiting Food and Drug Administration approval; U.S. introduction of Formulaid for pre-term infant formula is expected in 1996 and for all-infant formula in 1997.

Scratch-Resistant Lenses

Manufactured and marketed by Bausch & Lomb, Inc., Rochester, New York, the Ray-Ban® Survivors® Collection sunglasses feature a technological advance that makes them 10 times more scratch-resistant than conventional glass lenses.

The DiamondHard® technology involves coating the lenses with a film of diamond-like carbon (DLC) that not only provides scratch-protection but additionally reduces surface friction, so that the lenses shed water more easily to reduce spotting. The film coatings are supplied by Diamonex Optical Products Group, Allentown, Pennsylvania, which employs a modified version of a dual ion beam bonding process originally developed by Lewis Research Center.

The hardest substance known, diamond offers a wide range of potential applications but the potential was slow to develop because of the high cost. Interested in the possibilities of synthetic diamond coatings for aerospace systems, Lewis Research Center sought to get the advantages of diamond without the cost penalty by depositing a thin film of DLC on an inexpensive substrate (supporting material). Lewis conducted extensive research on the properties of DLC and ways to deposit the film on different types of substrates.

Among the coating methods developed was a technique known as direct ion deposition, in which an ion generator creates a stream of ions from a hydrocarbon gas source; the carbon ions impinge directly on the target substrate and "grow" into a thin DLC film.

Lewis patented the technology and subsequently licensed it to Air Products and Chemicals, Inc., Allentown, Pennsylvania, which was exploring aerospace applications of diamond coatings. An Air Products spinoff company—Diamonex—used the NASA technology along with its own proprietary technology in developing both polycrystalline diamond and DLC coatings for commercial optical products.

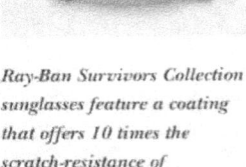

®Ray-Ban, Survivors and DiamondHard are registered trademarks of Bausch & Lomb, Inc.

Ray-Ban Survivors Collection sunglasses feature a coating that offers 10 times the scratch-resistance of conventional glass lenses.

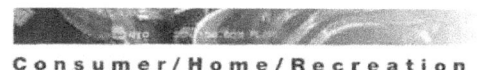

Miniature Earthmover

In 1995, International Machinery Corporation (IMC), Mt. Juliet, Tennessee completed the research and development phase of its first product—a miniaturized, remotely-controlled earthmover—and embarked on preparations to begin producing the machine late in 1996. The product is a 1/8 scale fully functional replica of the world's largest earthmover, the 220,000-pound Caterpillar® D11N Track-type Tractor. Caterpillar Inc., Peoria, Illinois granted IMC trademark product approval and a worldwide license to manufacture and market 1/8 scale replicas of Caterpillar products.

The Cat® D11N weighs only 340 pounds, has miniature hydraulic and transmission systems, and is powered by an IMC-designed seven horsepower, V4 four-stroke cycle engine which, says IMC, is the world's first miniature industrial internal combustion engine. The 45-pound engine, which will power all IMC products, is smaller, quieter and has less vibration than single cylinder engines of comparable power output.

IMC contemplates marketing the miniature earthmover to users of heavy equipment, particularly Cat equipment, such as fleet owners/contractors and operators of equipment for mining, construction and demolition work. The company sees broad potential for sales of the engine to Original Equipment Manufacturers for such applications as boat propulsion, light landscaping machinery, snow grooming vehicles, powered scale model vehicles, go carts and transportation for the handicapped.

A market area of particular interest to IMC is the special equipment category wherein the vehicle's small size, miniaturized water-cooled engine and remote control capability permit its use in confined spaces and in environments considered hazardous to humans. Prospective customers in this category include the military services and other federal agencies, law enforcement organizations, and waste management firms.

The initial products are operated by radio remote control. However, the company envisions large-scale export sales of the 1/8 scale Cat D11N and subsequent products; because of varying restrictions on radio frequency use among foreign countries, IMC is developing infrared remote control capabilities with technical assistance from NASA under multiple Technology Transfer Agreements; Marshall Space Flight Center is the cognizant NASA organization.

IMC plans to build upon the 1/8 scale Cat D11N experience by introducing other innovative systems, both Cat machine replicas and non-Cat products. Cat-licensed replicas planned for early introduction include the 994 Front-Wheel Loader and the 793 Off-Road Mining Truck. Additionally, IMC has preliminary designs for eight non-Cat products for military use in weapon systems, surveillance systems and explosive ordnance disposal.

®Caterpillar and Cat are registered trademarks of Caterpillar Inc.

Ribbed Swimsuit

Invented by Langley Research Center in the early 1980s, riblets are small, barely visible grooves that can be placed on the surface of an airplane to reduce skin friction—hence reduce aerodynamic drag—by modifying the turbulent airflow next to the skin. Although the grooves are no deeper than a scratch, they have a surprisingly beneficial influence on the airflow near the surface.

Riblets have found a wide range of non-aerospace applications, for example, they can be used to reduce friction or drag inside pipes and ducts, contributing to increased efficiency of pumps, heat exchangers and air conditioners. They have been used to good effect on eight-oared shells in regatta competitions and in America's Cup races (the *Stars and Stripes* racing yacht that brought the Cup back to the U.S. in 1987 had a hull whose underside was coated with riblets).

A new application of riblets has appeared in the Strush SR® competition swimsuit, marketed by Arena North America, Englewood, Colorado. Arena, which is the official supplier of apparel to the Triathlon Federation U.S.A., the national organizing committee of triathlons, combined the Langley riblet technology with company-developed innovations to produce a swimsuit that, says Arena, "has been flume tested to be 10 to 15 percent faster than any other world class swimsuit."

The Strush SR design employs a silicon ribbing in the areas of the swimsuit (chest and buttocks) subject to the most turbulence in the water, reducing hydrodynamic resistance. In addition to the ribbing, Arena uses micro fibers and special treatments to reduce the amount of water absorbed by the suit.

The Strush SR swimsuits are "stroke specific," meaning that they are constructed differently to maximize the effectiveness of the silicon ribbing for each of four different stroke disciplines: freestyle, backstroke, butterfly and breaststroke. The stroke specific swimsuits made their debut in competition in 1995, at the Pan American Games in Mar Del Plata, Argentina. The results were impressive: swimmers wearing Strush SR suits won 13 gold medals, three silver and one bronze.

®Strush SR is a registered trademark of Arena North America.

The Strush SR competition swimsuit employs NASA-developed riblets to reduce water resistance.

Television Transmission Technology

A spinoff system that offers a substantial improvement in UHF-TV transmission efficiency is the VKP-7990 MDC Klystron, which is manufactured by the Microwave Power Tube Division of Communications and Power Industries, (CPI), Palo Alto, California. There are 90 CPI klystrons in operational service at 36 UHF-TV stations.

The MDC klystron is the result of a multiyear cooperative development program by a group that included Lewis Research Center, the National Association of Broadcasters, the Public Broadcasting System, TV transmitter manufacturers, and Varian Associates, Inc., also of Palo Alto. Throughout the klystron development period, the Microwave Power Tube Division was part of Varian, but Varian's Electron Devices Group has since separated from Varian to become CPI.

The program was initiated to address a problem experienced by the UHF segment of the TV industry: for adequate reception, UHF stations need greater transmitter power than their VHF counterparts and, additionally, UHF transmitters are inherently less efficient; thus, UHF station operators had to pay substantially higher electric utility costs, a significant competitive disadvantage.

The development group sought a way to make available to UHF operators power amplifying devices with efficiencies comparable to VHF. The indicated line of approach was to incorporate into UHF transmitters a power amplifying device known as the Multistage Depressed Collector (MDC) developed a decade earlier at Lewis Research Center to enhance the efficiency of communications satellite transmissions; the MDC allowed satellites to transmit more powerful signals, thus enabling the use of smaller, less costly Earth stations for signal reception.

The klystron is a vacuum tube used to generate and amplify ultrahigh frequencies. It draws radio frequency energy from a high voltage electron beam but does so at very low effi-

ciency levels; most of the energy is dissipated as waste heat. The concept behind the Lewis/Varian development was that the MDC could recover much of the wasted energy by recycling a large part of the electron beam energy, in effect doubling the amount of the beam energy being converted to radio frequency energy.

The program began in 1984. Within the initially targeted three years, a successful MDC klystron had been produced by Varian and thoroughly tested. It took another two years to integrate the klystron into a complete transmitter for field operation. Varian started commercial production of the tubes in 1990 and CPI took it over in August 1995.

CPI continues to monitor the klystron's performance. According to CPI senior scientist Earl W. McCune, the total operating time for the tubes in service has exceeded three million hours; projected Mean Time Between Failures is 60,000 hours; the klystrons show no significant adverse effects due to incorporation of the MDC feature; and the initial project goal of cutting electric power consumption by half in UHF-TV transmitters has been realized.

The CPI MDC klystron cuts the electric power consumption of UHF-TV transmitters by half.

Sensors for Environmental Control

An advanced system for monitoring industrial process fluids highlights spinoffs in environment and resources management

A particularly productive source of spinoff products and processes is the Small Business Innovation Research (SBIR) program, which was established by Congress with two objectives: to increase participation of small businesses in federal high technology R&D activities, and to stimulate conversion of government-funded research into commercial applications.

NASA's SBIR program has been eminently successful. It provides NASA an additional source—beyond traditional aerospace companies—of R&D talent and innovative thought. Hundreds of new systems and components that advance NASA's capabilities have emerged from the SBIR program. On the other side of the coin, NASA SBIR contracts give small businesses opportunities to hone their R&D skills and expand their technological capabilities, and in many cases an SBIR assignment leads to a contractor's development, with private capital, of a commercially useful product or process.

An example of environment-related research that led to commercialization of new types of sensors is the SBIR collaboration of Kennedy Space Center (KSC) with GEO-CENTERS, Inc., Newton Centre, Massachusetts. GEO-CENTERS is a company dedicated to basic and application-oriented R&D over a broad spectrum of technological disciplines. The company has special capabilities in designing fiber optic sensors for such applications as measurement and characterization of industrial processes, remote sensing of chemical/biological hazards, and environmental monitoring.

KSC's initial SBIR contract with GEO-CENTERS (1990) sought development of an instrument for use in space life support research (investigation of the air/water recycling potential of certain plants). An important element of such research is accurate measurement of the hydroponic culture's pH factor, an indication of the acidity or alkalinity level of a solution. NASA was interested in an advanced type of sensor known as an "optrode," the optical-based equivalent of the glass pH electrode. A series of optrodes would provide information about maintaining the proper conditions for the plants' survival in a "space greenhouse."

Over a two-year span, GEO-CENTERS successfully developed an optrode for NASA use and, since the instrument clearly had strong commercial potential, embarked on development of a PC Based pH Monitoring System for industrial use.

Now commercially available, the system employs a proprietary fiber optic sensor (the optrode) to enable long term continuous monitoring of the pH level of fluids in both standing and flowing conditions. In addition to the optrode, the system includes an opto-electronic (printed circuit) board, equipped with light sensors and detectors, that fits into a standard desktop computer; and a fiber optic cable connecting the circuit board and optrode through which communications (light signals) are relayed. The optoelectronics board supplies energy to the optrode and converts the optrode's optical signals to electrical signals. System software provides pH display, data logging, calibration and diagnostic functions.

A typical application is monitoring the pH of an industrial process stream to assure that the pH values of the effluent discharged from the plant are within the acceptable range specified by the Environmental Protection Agency. Normally, groundwater has a pH of 5.5 to 7.5; when used in chemical processing, the pH level can fall outside the normal limits and become harmful to fish and wildlife. Therefore, the pH level of a process stream must be continuously monitored to verify acceptability and, if it is not in compliance, it must be restored to the acceptable range.

GEO-CENTERS literature cites a number of advantages of the fiber optic pH monitor: low maintenance, long term calibration stability, high resistance to biological/chemical fouling, and immunity to electromagnetic interference

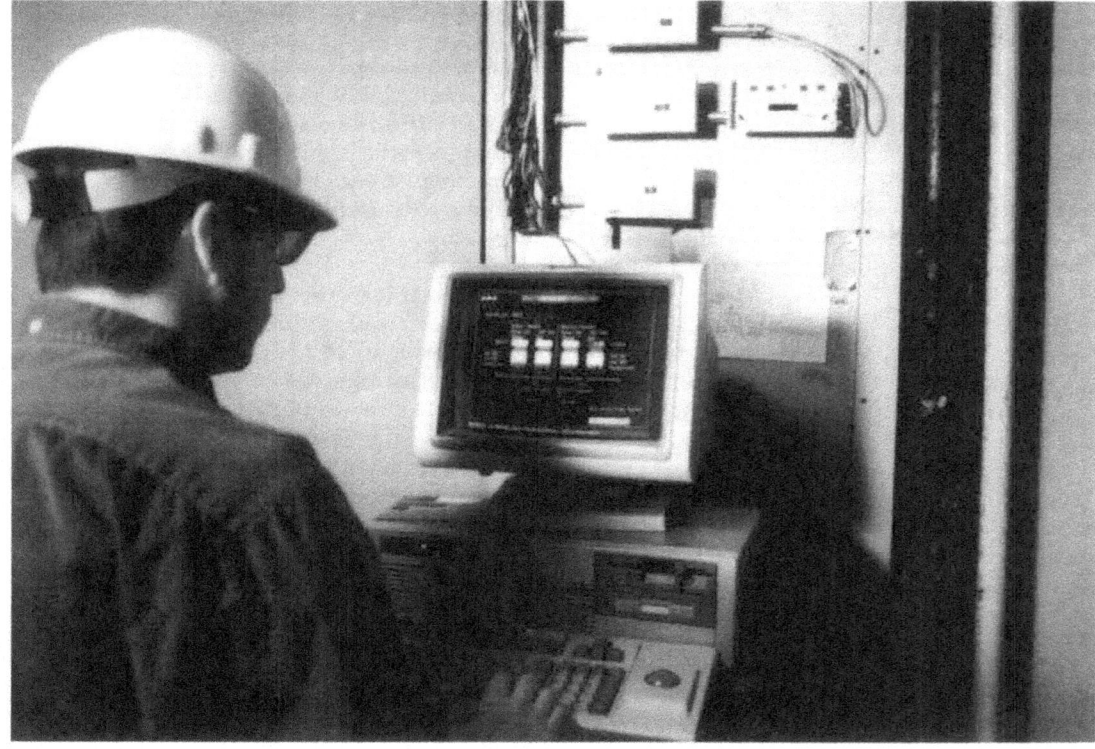

A plant technician monitors an industrial process stream to assure that the effluent water's pH level is in compliance with environmental regulations. Recently introduced to the commercial market by GEO-CENTERS, Inc. the pH sensing system exemplifies the broad range of spinoff products emerging from the NASA Small Business Innovation Research program.

caused by radio waves, lightning, motors or generators, because the system is optically based rather than electrically based. In addition, the optrode can operate in solutions with very low conductivity, an important factor in the measurement of very clean—hence low conductivity—boiler water at power plants.

In 1993, when development of the commercial pH sensor was in progress, KSC awarded GEO-CENTERS another SBIR contract related to space recycling research, this one for a sensor capable of continuously monitoring dissolved ammonia in a bioreactor and an aquaculture. This development was successfully accomplished in 1995 and commercialized with the first deliveries late in that year.

The ammonia sensor employs the same general configuration as the pH sensor, including a replaceable sensing element and the computer-compatible opto-electronic board. The application, however, is different:

it is intended for monitoring bioreactors rather than process streams. Bioreactors are complex systems containing biological components (bacteria, yeast, mammalian cells). Used to synthesize compounds, such as pharmaceuticals, bioreactors must maintain colonies of living organisms whose density and health determine the ultimate productivity of the reactor. The ammonia concentration—along with pH—is one of the important variables that must be controlled to assure proper conditions for maintaining a colony. There is, says GEO-CENTERS, no other means available for continuously monitoring ammonia levels in a bioreactor and the company sees a significant potential market for its ammonia sensor.

Wind Monitor

Orbital Sciences Corporation (OSC), Fairfax, Virginia is a world leader in the design and manufacture of small space systems intended to reduce the cost of space operations and thereby broaden the market for space products and services.

Among the company's extensive line of products is a family of inflatable structures for the atmospheric monitoring community. One such inflatable, known as the "Jimsphere," is of particular interest because, although developed more than 30 years ago, it remains the standard at all U.S. missile/launch vehicle ranges for obtaining accurate upper level wind data.

The Jimsphere wind measurement balloon is made of lightweight radar-reflective materials.

Large missiles and launch vehicles are very sensitive to windshear, especially as they approach the area of maximum dynamic pressure, typically between 30,000 and 50,000 feet. Prior to every flight today, launch teams analyze a computer-generated flight profile that involves detailed specifications of the wind field through which the vehicle must fly. In the early 1960s, however, no method existed for making high resolution measurements of the wind profile. At that time NASA was already developing the Saturn launch vehicles for the Apollo lunar landing program and it was essential that NASA also develop a meteorological sensor of superior aerodynamic stability to determine the vertical gradients of the wind before Saturn launches commenced.

The standard smooth-surface weather balloon could not do the job. The reason: the smooth balloon was subject to zigzagging or spiraling as it ascended, due to large air vortices that shed off the surface at various positions; this caused sporadic horizontal motions of the rising balloon that made accurate radar-tracking measurement of the balloon impossible.

After several NASA-sponsored studies failed to provide a suitable method, a NASA engineer came up with an answer. Dr. James R. Scoggins, today director of meteorological studies at Texas A&M University, then at Marshall Space Flight Center, took a simple approach to a complex problem: rather than invent a new system, change the characteristics of the existing system, the smooth surface balloon. He bought some conical dixie cups and attached them to a balloon to "rough up" the smooth surface. The cones, or "roughness element," were intended to prevent the formation of vortices and thus damp the sporadic motions; additionally they increased drag. The combination of reduced lift and increased drag stabilized the balloon so that, when it entered a changing wind field, it would quickly assume the speed of the wind without zigzagging. It worked; named the Jimsphere in honor of its inventor, the system was assigned to a company for refinement and production and OSC subsequently acquired the patent rights.

The Jimsphere now being produced is a balloon two meters in diameter, made of lightweight, flexible, radar-reflective materials. Ground radar, navaids or theodolites track position and collect wind data from the balloon at altitudes up to about 10 miles. Jimsphere data was used in the design of the Saturn vehicles, the Space Shuttle and other launch systems. The balloon has supplied pre-launch wind data for all NASA/Air Force ground-based rocket launches from Cape Kennedy, Kennedy Space Center and Vandenberg Air Force Base.

Mower/Litter Remover

An example of how NASA technical assistance can solve developmental problems encountered by industrial firms and offer significant improvement to the end product is contained in the experience of The Burg Corporation, Sylacauga, Alabama.

The Burg Corporation manufactures the Vac-N-Bag, a tractor-drawn combination grass mower/litter remover for manicuring lawns, parks areas, highway medians and other areas. On the same pass, the machine cuts the grass, collects the grass cuttings and sucks up all litter in the passed-over area. The Vac-N-Bag can operate as a litter gatherer on dirt, asphalt and concrete, and it runs forward and backward (no need to turn around in tight places). It has a generously-sized refuse bag that is small enough for single-operator handling and Burg also offers an optional large container for leaf collection.

In developing the prototype machine, Burg produced a workable vehicle but company officials sought a way of getting more power out of the suction system. Company president Cecil Thornburg sought help from Marshall Space Flight Center (MSFC) by submitting a problem statement and request for assistance. Engineers of MSFC's Turbomachinery Branch reviewed the Vac-N-Bag's suction system and devised a way of guiding heavier items of trash to a point where suction was greatest, thereby insuring trash pickup. MSFC also suggested changes to the impeller and the exhaust port.

The MSFC recommendations were enthusiastically accepted and incorporated in a modified version of the Vac-N-Bag, and Burg's Cecil Thornburg reported to MSFC a "vast improvement" in the machine's performance. The modifications were adopted for the Burg Model II Vac-N-Bag, now in production and in service with a number of highway departments, city governments and park authorities. One user—Childersburg (Alabama) Department of Parks and Recreation—reported a halving of work time by vacuuming trash and clippings while cutting grass.

A Burg Corporation employee demonstrates the Vac-N-Bag, a machine that sucks up litter while mowing grass.

An accessory allows hosing up debris from gutters, shrubbery and picnic areas.

Stress-detection Lenses

NASA and other organizations have conducted extensive research on early detection of stress in vegetation through use of various sensory devices that measure radiations emitted or reflected by plants. An extension of that technology is a simple, inexpensive home garden variety plant stress detector, a pair of rose-tinted lenses that enable a viewer to see otherwise invisible signs of plant deterioration. Known as Hawkeye Lenses, the glasses are marketed by Optical Sales Corporation, Portland, Oregon; they incorporate technology developed by a NASA scientist—Dr. Leonard A. Haslim of Ames Research Center—and they were introduced to the commercial market through the technology transfer efforts of the Ames Commercial Technology Office.

The Hawkeye lenses serve as a "passive chlorophyll detector." When drought, pests, disease or other agents cause stress to a plant by reducing chlorophyll, the initial plant damage is not visible to the naked eye except under a microscope. The lenses make it visible by means of dyes, integrated into the lenses, that filter out certain wavelengths of light.

When one looks at a green leaf through a Hawkeye lens, the stressed part of the leaf is visually enhanced and it stands out from the healthy part of the plant by appearing pink or brownish red. "The filtering lenses provide a color discrimination due to the way plants absorb and reflect sunlight," explains Dr. Robert Brock, president of Optical Sales Corporation.

The glasses sell for less than $100 and can be made into prescription glasses. "The glasses work because they enhance the edges of the visual spectrum, allowing the user to see what normally can't be seen," Brock adds. "Plant stress becomes visible earlier, at a critical stage when something can often be done."

Optical Sales Corporation is continuing stress detection research under a Department of Agriculture contract; among the researchers is Dr. Carlos Blazquez of the University of Florida, who has worked with NASA in infrared detection of stress in Florida's citrus groves. The advanced phase of the research is aimed at refining the technology and tailoring it to match the light spectra reflected by various specific plant types.

Hawkeye lenses enable a viewer to see otherwise invisible signs of plant deterioration.

Turfgrass Conditioner

Plant-Wise Biostimulant Company, Louisville, Kentucky, a relatively new company formed in 1992, is marketing a novel product for the turfgrass professional that is designed to improve turf quality and vigor even under adverse conditions. Called 3D Concentrated Plant Growth Supplement, the product's formula incorporates space agricultural technology developed under NASA contract.

3D, says the company, is a scientifically balanced blend of fortified seaweed extracts, humic acid and plant nutrients designed to supply turf plants with extra "insurance" to handle stress-related problems and maximize their growth potential. The "3D" refers to the product's three principal conditioning characteristics: foliar enhancement, the main benefits of which are improved color and aesthetic quality, reduced stress and frost damage; physiological integrity, wherein the plant's biochemistry is improved by formulation components that initiate cell division, which helps the plant stimulate development of new leaves; and foundation fortification, promoted by components that help the plant generate new and deeper roots while retarding root senescence (loss of vigor).

Field tested by Dr. Dick Schmidt, professor of turf ecology, and his staff at Virginia Polytechnic Institute, Blacksburg, Virginia, 3D won a solid endorsement. Dr. Schmidt reported: "With the use of 3D, a balanced blend of biostimulant and iron, we have demonstrated that turf quality can be enhanced by reducing the influence of drought, increasing root development under adverse conditions, and offsetting the infection of certain diseases and nematodes.

The 3D formulation was developed by Dr. Richard R. Dedolph, formerly a plant physiologist with Argonne National Laboratories and Western Regional Research Laboratories, who also worked under NASA contract as a project leader in research on plant cultural systems for space application. The 3D development benefitted from Dr. Dedolph's NASA-acquired expertise in research that explored the potential of plants as food sources and recycling agents in long duration spacecraft.

The lush green and fairway turf at Pennsylvania's Butter Valley Golf Port is enhanced by 3D Concentrated Plant Growth Supplement.

Hazardous Environment Robotics

Deneb Robotics, Inc., Auburn Hills, Michigan is an internationally known leader in 3D graphics-based factory simulation, telerobotics and virtual reality software used widely in the aerospace, automotive, defense, environmental, medical, nuclear and research communities.

Among the company's broad software product line is TELEGRIP™, which provides a 3D graphical interface for previewing, interactive programming and real-time bilateral control of remote robotic devices. It provides operators a system for safe, quick and efficient remediation of hazardous environments from a single point of control and input that is isolated from virtually all operator hazards.

Accurate 3D kinematic models of the robot and work space components allow the operator to preplan and optimize robot trajectories before the program is automatically generated. TELEGRIP can verify all geometries and dynamically update the model to adjust the environment in real-time. Control commands are monitored when running in autonomous, teleoperational or shared control modes to assure procedural safety.

A key feature of TELEGRIP is a Video Overlay Option that utilizes video to calibrate 3D computer models with the actual environment. The video overlay technique is especially useful for on-line planning applications or teleoperations in remote, hazardous or complex environments such as space, undersea or nuclear sites. In such environments specific

Dr. Won Soo Kim, TELEGRIP inventor, remotely operates a robot.

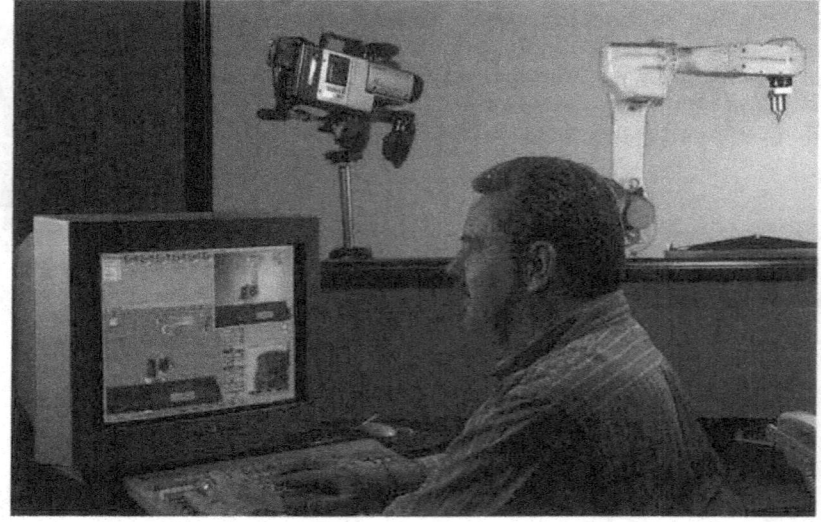

A video camera provides a Deneb engineer with a view of the robot.

conditions may be unknown, or they may change suddenly while robotic operations are in progress; the advanced control capabilities provided by TELEGRIP enable ready adjustment to change.

The Video Overlay Option is the result of a technology sharing agreement with NASA's Jet Propulsion Laboratory (JPL). JPL developed a virtual reality calibration technique for reliable and accurate matching of a graphically-simulated environment in 3D geometry with actual video camera views. The system was designed for predictive displays with calibrated graphics that overlay in live video for telerobotics applications, for example, the system allows an operator to designate precise movements of a robot arm before sending the command to execute.

Following successful test of the video overlay techniques, JPL concluded a technology cooperation agreement with Deneb Robotics that allows the company to integrate Video Overlay into the commercially available TELEGRIP to expand its utility in hazardous environment robotics.

™TELEGRIP is a trademark of Deneb Robotics, Inc.

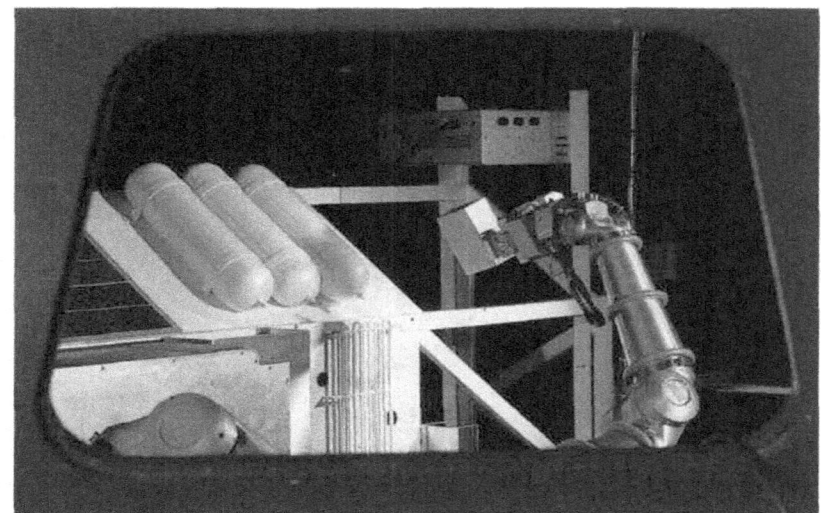

By selecting critical control points of reference, the operator and robot create the robot's work cell movement paths.

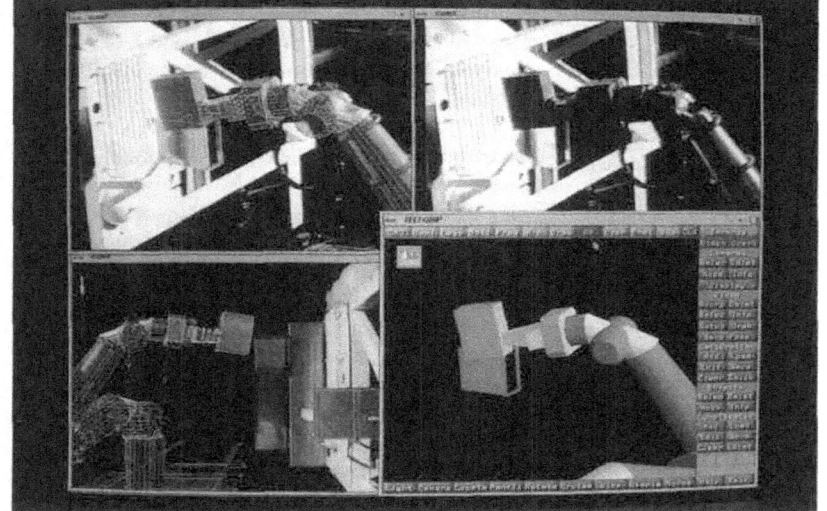

The operator can view the video image of the real world environment (upper right) and the computer's interpretation of the same scene using TELEGRIP.

Telemetry Systems

TSI/TelSys Inc., Columbia, Maryland, the American subsidiary of TSI/TelSys Corporation, Vancouver, British Columbia, Canada, is a spinoff company formed expressly to commercialize NASA high-data-rate telemetry technology and products originally developed at Goddard Space Flight Center's Microelectronic Systems Branch. The company exemplifies two different technology transfer routes: the growing process of "privatizing" certain types of government operations, and the personnel technology transfer medium, whereby NASA employees leave the agency to join private industry and use their NASA acquired technological expertise to develop products for the commercial market.

TelSys develops and manufactures ground station communications equipment that performs both traditional telemetry processing and the bridging/switching functions required to interconnect local/wide area networks with space-ground communications networks.

The company continues to support government sponsored space programs, such as those operated by NASA, the Department of Defense and the European Space Agency. In addition, however, TelSys is moving more and more into commercial communications spheres. The remote sensing industry is experiencing substantial activity growth and TelSys is manufacturing systems that provide affordable, timely access to space-acquired data in both scientific and commercial remote sensing programs. Although the company's products and technology were developed for satellite telemetry applications, the existing network interfaces and the inherent ability of the equipment to process high-rate streams of multimedia (voice, imagery, text) data enable effective interconnection of broadband networks using commercial communications satellites.

A spinoff company formed to commercialize NASA high-data-rate telemetry technology, TSI/TelSys manufactures a high speed processing system for commercial communications applications.

The sequence of events that led to the formation of TelSys began in 1985, when NASA tasked James Chesney, then a 16-year veteran of the agency, with developing technology to support the next generation of ground telemetry systems. The challenge was to develop systems capable of meeting NASA's increasingly sophisticated requirements by processing data at rates up to 300 million bits per second, and to develop maximal commonality, reusability and the interoperability among all new systems deployed by NASA.

In 1994, James Chesney retired from NASA and founded TSI/TelSys Inc. In 1995, he was joined by other former members of the

Goddard Microelectronic Systems Branch and by a number of key hires from industry.

Today, TSI/TelSys Inc. designs, manufacturers, markets and supports a broad range of commercial satellite telecommunications gateway products evolved from technologies that originated in the Microelectronics Systems Branch. These technologies and products support two-way, high-speed space data communications for telemetry, satellite remote sensing, and high-data- rate communications applications.

A unique high performance chip, developed by the TelSys group when it was a NASA unit, is produced in units of 24.

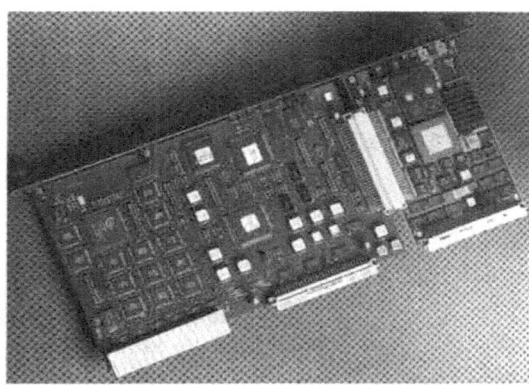

The specially-designed chip is incorporated into TelSys product boards to filter unwanted information from data transmitted by satellites to Earth.

The satellite antenna is part of the system used for high-speed data transmissions.

Air Quality Monitor

Highlighting

spinoff examples

in the field of

computer

technology is

a system

for analyzing

smokestack

emissions

Increasingly stringent government regulations for control of the environment require industrial firms to install increasingly sophisticated monitoring equipment for full compliance with smokestack emission standards. One such advanced system is the Stak-Tracker™ CEM (Continuous Emission Monitor) Gas Analyzer, an air quality monitor capable of separating the various gases in a bulk exhaust stream and determining the amounts of individual gases present within the stream.

The Stak-Tracker is produced by GE Reuter-Stokes, Twinsburg, Ohio, a subsidiary of General Electric Company, and is supported by the GE Corporate Research & Development Center (GE CR&D), Schenectady, New York. An important element of the Stak-Tracker is a NASA-developed software package that made possible the system's advanced analytical technique. This technique is the key to accurate measurement of minute quantities of certain gases, such as nitric oxide, carbon monoxide and sulphur dioxide, generated in new, low-emission combustors.

Developed by Langley Research Center, the method of analysis is a gas filter correlation technique; it measures the concentration of an individual target gas within the bulk stream by determining the degree to which molecules of that gas absorb an infrared beam.

The Stak-Tracker directs an infrared beam across a smokestack exhaust to a reflector and collects the reflected beam. The instrument sends the beam through a dual filter assembly. The first filter operates in the general wavelength range of the target gas. The other filter, which oscillates in and out of the infrared beam, is the target gas cell. When the cell filter is in the beam, the beam is absorbed by the cell at the target gas' specific wavelength before it goes through the smokestack exhaust. A detector measures the ratio of the two signals, which serves as the basic input that enables the patented analysis technique to provide a measurement of the target gas concentration. The Stak-Tracker can measure up to six gas components within three seconds.

In the early 1990s, when the GE CR&D was helping GE Reuter-Stokes develop the system, it became apparent that there were no commercially-available software tools for modeling

Right: A pair of industrial smokestacks equipped with GE Reuter-Stokes Stak-Tracker gas analyzers; the Stak-Tracker equipment is shown in closeup (far right). Offering highly accurate measurements of pollutants for improved compliance with environmental regulations, the system incorporates software technology that originated at Langley Research Center.

the performance of sensitive pollution monitors of this type. The developers, however, were aware that Langley Research Center had designed a software system for a Gas Filter Correlation Radiometer (GFCR) that seemed the answer to the need.

Two GE CR&D scientists—Dr. Emily Shu of the Industrial Electronics Laboratory and Dr. M. K. Cueman of the Manufacturing Technology Laboratory—visited Langley's Atmospheric Research Division and received detailed information on the GFCR software, which is capable of calculating gas absorption even when the gas is present in minute amounts, and which additionally can separate the interference from a number of coexisting gases in a stream. The visit led to a GE-NASA Space Act Agreement for exchange of technical information relative to trace gas measurement. Langley scientists shared their expertise in atmospheric research and helped GE apply the

GFCR code. The GFCR software was incorporated into the calibration of the Stak-Tracker, and the software has found additional use at GE CR&D for evaluating changes in the instrument's design and solving other industrial problems.

Besides its utility as an aid to full compliance with environmental regulations, the Stak-Tracker offers fast response for process control applications and relatively low installation and maintenance costs. It is applicable to gas turbines, gas, oil, or coal-fired boilers, incinerators, dryers, scrubber controls, kilns and process heaters used by the power, oil, pulp and paper, iron and steel, non-ferrous metals, cement, glass and other industries.

[1]Stak-Tracker is a trademark of General Electric Company

Video Compression

The OPTIVideo ™ MPEG Encoder and Decoder are two members of a family of products developed by Optivision, Inc., Palo Alto, California to speed up audio/video processing time and reduce costs. The encoder/decoder products were spawned by a NASA Small Business Innovation Research (SBIR) project.

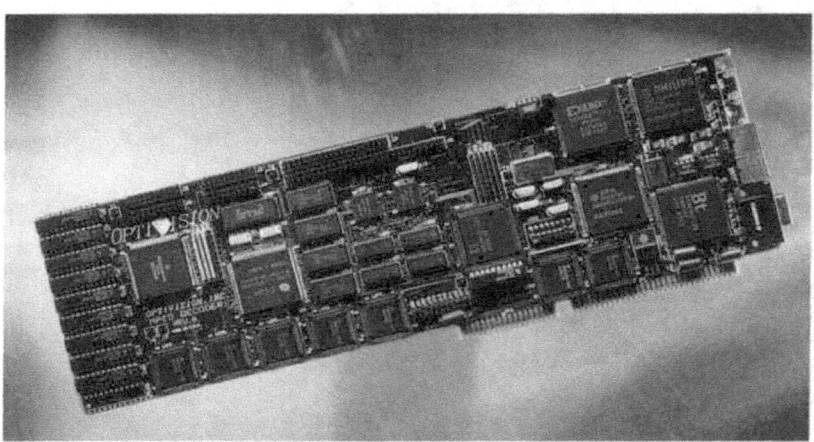

The OPTIVideo Encoder converts video tapes and discs to compressed digital form at 30 frames a second.

The companion OPTIVideo Decoder decompresses the bit stream to provide high quality digital playback.

Established by Congress in 1982, the SBIR program has two major objectives: to increase participation of small businesses in federal R&D activities, and to stimulate conversion of government-funded research into practical products for the commercial market.

The commercial OPTIVideo systems stemmed from a Goddard Space Flight Center SBIR contract awarded to Optivision for development of two PC-compatible boards (the encoder and decoder) and associated software for realtime video compression and decompression; the research was intended to support NASA applications in such areas as telerobotics, telesciences and spaceborne experimentation.

While working on the Goddard contract, Optivision committed its own internal funding to parallel development of the two commercial products. Both the government and commercial efforts proved successful, and Optivision introduced its first commercial MPEG encoders/decoders in 1993-94. The company has since sold more than 600 systems to customers in the telecommunications, cable, broadcast TV and CD-ROM markets. They are used in such applications as television advertisement insertion; video CD authoring; interactive video databases; video transmission; remote learning; and video-on-demand.

The systems offer realtime encoding and decoding at 30 frames a second. The MPEG encoder converts video source material, such as VHS tapes and laser discs, to a manageable compressed digital form that can be easily stored or transmitted (the MPEG refers to the compression algorithm, the standard established by the international Motion Picture Expert Group). The decoder decompresses bit streams to provide high quality digital playback of full screen video and CD quality stereo audio.

™OPTIVideo is a trademark of Optivision, Inc.

Intelligent Agent Technology

A problem computer users face is that as applications become more powerful, they also become more complex. And while users cope with the complexity, they must still deal with the mundane aspects of computer usage, such as backing up, repetitive file management and arranging windows. Charles River Analytics, Inc., Cambridge, Massachusetts has brought to the market an autonomous intelligent software product that relieves users of many routine housekeeping and preparation tasks.

Called Open Sesame!, the software "learns" a user's behavior and offers automation and coaching suggestions to the user; it reduces the number of mouse and keyboard operations required to accomplish a given job because it automatically carries out actions that were previously done manually. The product's developers credit NASA funding support and technology background with "critical" assistance in developing Open Sesame! and other company software components.

Available for Apple Macintosh computers, Open Sesame! is based on Charles River Analytics' hybrid neural network/expert system technology. It compares high level events (like opening a folder or quitting an application), generated by the user's mouse clicks and key strokes, to information stored in its neural learning module and its inference engine. The neural learning module looks for repetitive patterns that have not been automated; when it finds one, it creates an observation and, on approval, automates the task it observed.

Founded in 1983, Charles River Analytics has both personnel expertise and technical roots in NASA technology. Prior to starting the company, president Dr. Alper Caglayan had worked five years at Langley Research Center on digital flight systems R&D. Dr. Greg Zacharias, vice president, served at Johnson Space Center, where he worked on the preliminary design of the Space Shuttle Orbiter's reentry autopilot.

In addition to this NASA-acquired knowledge base, much of the company's computational intelligence technology was developed under NASA Small Business Innovation Research (SBIR) grants. A Langley SBIR project on neural guidance research generated software components later incorporated in Open Sesame! A Johnson-sponsored SBIR contact on hybrid neural network expert systems environment was instrumental in development of another Charles River product, a neural expert software system known as NeuX.

Open Sesame! software reduces computer-use complexity by relieving users of many routine tasks.

Semiconductor Cubing

A Memory Short Stack™ is a three-dimensional semiconductor package in which dozens of integrated circuits are stacked one atop another to form a cube. Known as "chip-stacking" or "cubing," the innovative process was invented by Irvine Sensors Corporation (ISC), Costa Mesa, California from technology developed under NASA and Department of Defense contracts. The cubing technique offers faster processing speeds, higher levels of integration and lower power requirements than conventional chip sets; it also offers dramatic reduction in the size and weight of memory-intensive systems such as medical imaging devices.

ISC's cubing technology is regarded as a major advancement in high density computer electronics. The Memory Short Stack won a 1993 NASA Award of Innovation and in 1994 was selected by *R&D Magazine* for an R&D 100 Award, presented annually to the year's 100 most technologically significant new products.

*A Memory Short Stack
is a cube-shaped package
of integrated circuits.*

ISC initiated its stacking technology development in the early 1990s. Department of Defense contracts led to development of a "full stack" version of the technology, but the height of the full stack was too great for the multichip packages employed in airborne and spaceborne electronic systems, the primary market envisioned at that time. The company therefore embarked on development of the Memory Short Stack concept, which involved a more limited number of chips mounted like a stack of pancakes. This approach lowered the height of the stack, permitted its attachment using conventional surface-mounting techniques, and broadened the number of electronics packages in which it could be used.

The Memory Short Stack that evolved from the company's development effort is fabricated by vertical assembly of integrated circuits, thinned to as little as seven-thousandths of an inch, and extremely thin layers of laminate. This configuration significantly increases density and maximizes efficient use of design space. Typically, cubes average between 40 and 50 integrated circuits.

Although ISC originally focused on military and space applications of 3D chip-stacking, it is now introducing the technology to a broadening range of commercial market applications, including computers and telecommunications. The company has established a joint development program with IBM Corporation, Burlington, Vermont and is exploring other alliances with major computer manufacturers. Current ISC products include a four-megabit Wide Word Memory Short Stack and DRAM Memory Short Stacks in 64-, 80- and 160-megabit configurations.

Under the Small Business Innovation Research (SBIR) program, NASA has awarded Irvine Sensors eight Phase I and five Phase II contracts, including an ongoing project at Jet Propulsion Laboratory (JPL) to evaluate chip-stacking for space systems. The NASA contracts contributed importantly to the commercialization of the cubing technology. One of them, sponsored by Goddard Space Flight Center, involved application of the Memory Short

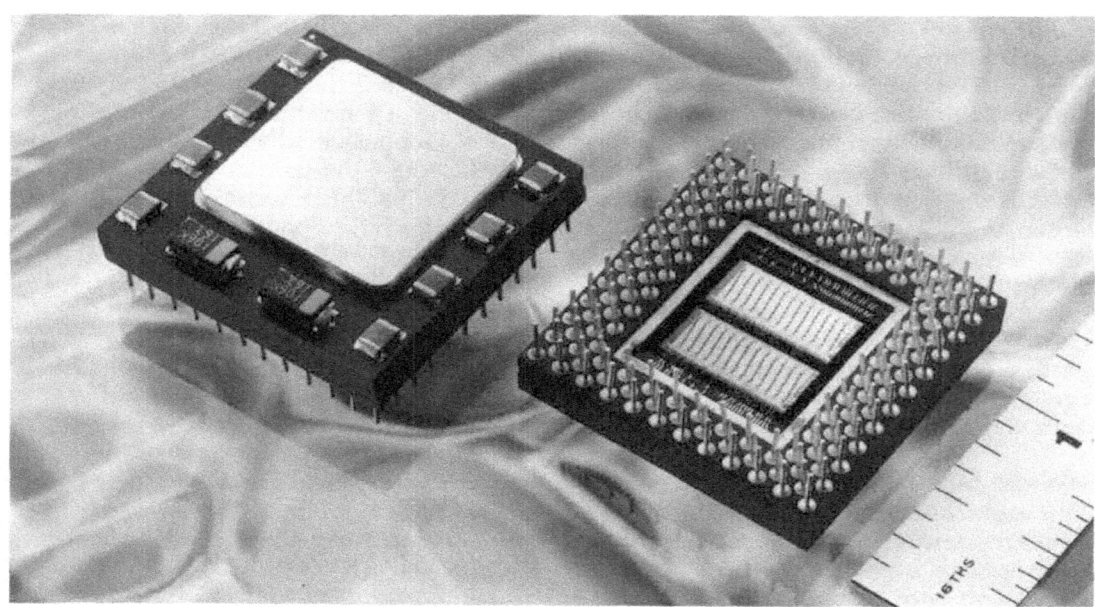

Stack to a spaceborne data recorder. That project significantly advanced the commercialization potential because it called for designing the short stacks so that they would be compatible with such industry standard attachment techniques as wire-bonding and tape-automated bonding; this configuration made Memory Short Stacks usable in virtually all types of electronic assemblies.

In addition, the Goddard SBIR attracted the attention of IBM Corporation and led to the ISC/IBM partnership designed to facilitate high volume, cost-competitive production of short stacks for the commercial market. ISC and IBM share a Cubing Process Development Center at Essex Junction, Vermont. Deliveries of the first Memory Short Stacks produced at Essex Junction began in 1994.

Another NASA contract of special importance is a JPL contract for a processing node known as SOBIEC (Spacecraft On Board Information Extraction Computer). This work takes the 3D stacking technology one step further by combining Memory Short Stacks with parallel processing. The commercial potential for this advancement of the technology is described as "enormous."

An advanced development for a NASA spacecraft computer combines 3D stacking with parallel processing.

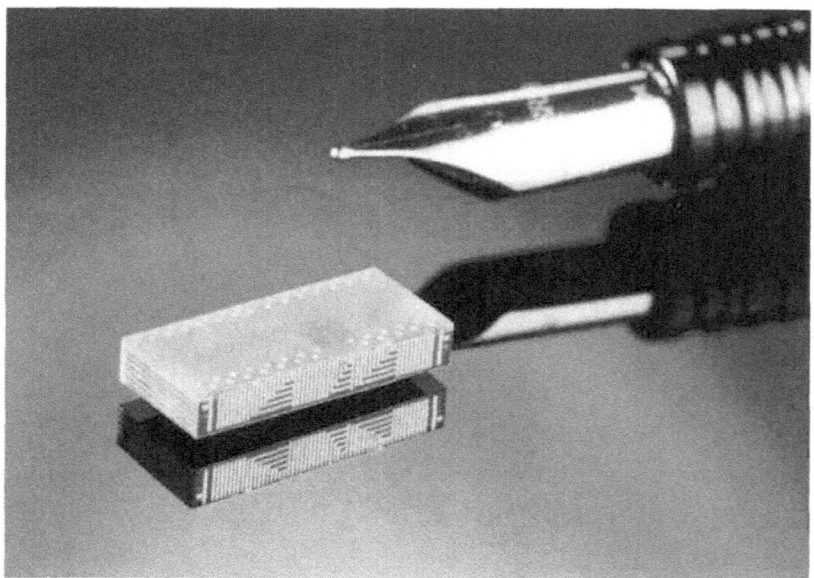

The DRAM 80-Megabit Short Stack exemplifies the small size of the stacked package.

™Memory Short Stack is a trademark of Irvine Sensors Corporation.

Simulation Software

NASA is one of the pioneering organizations advancing the art of computational mechanics, which involves creation of mathematical models of a product design and testing various configurations by computer simulation before settling on a final design. Initially used by NASA to model designs of flight vehicles, the technique has expanded enormously over the last two decades to embrace calculation and visual imagery of forces and phenomena that cannot be simulated in conventional facilities, such as wind tunnels. Its use has similarly grown from air and space applications to cover a great variety of non-aerospace products from autos to rollercoasters to medical equipment.

These advances have been made possible by a great wealth of software developed by government agencies and private companies. The Small Business Innovation Research (SBIR) program, through which NASA encourages commercialization of technologies that have

both public and private sector applications (see page 108), has been particularly productive of advanced software development. An example in the field of computational simulation is the work of Computational Mechanics Company, Inc. (COMCO), Austin, Texas. Over the past decade, COMCO has worked on several NASA SBIR contracts for computational mechanics modeling and simulation software, and a number of COMCO's products have successfully been commercialized.

The company's basic software is the PHLEX™ family of *hp*-adaptive finite element analysis programs, a library of object-based modules that may be linked and extended to create job-specific codes across a broad spectrum of engineering disciplines. NASA's research assistance was integral to the development of PHLEX, says COMCO; several NASA SBIRs allowed the company to improve and expand upon the PHLEX code and make it work for a

Workers at a Texaco oil field drill a vertical oil well. Texaco's oil production is aided by a computational simulation code called UNISIM.

A Texaco inspector performs a routine quality control check.

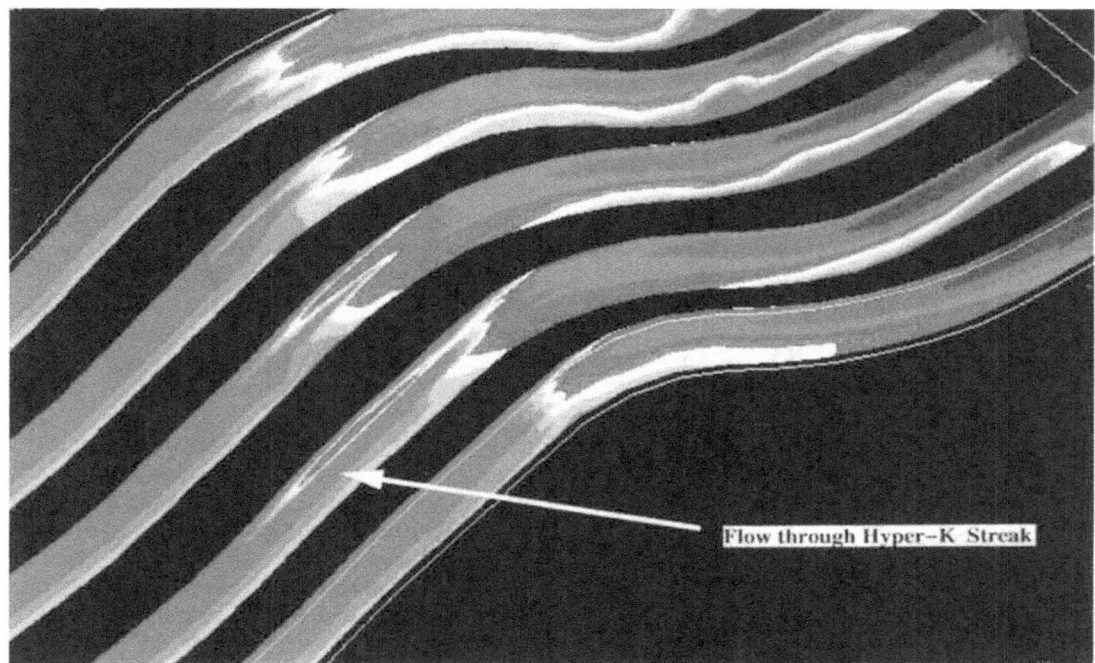

Flow through Hyper–K Streak

UNISIM simulation software produces imagery and data for pressure and saturation solutions.

variety of computational mechanics problems, including problems in both fluid and solid mechanics. Traditional general purpose finite element software requires the expertise of an analyst to appreciate the quality of the numerical simulations independently of the quality of the design. PHLEX-based software is designed to remove the uncertainty about the quality of the solution by estimating the solution's error and producing, automatically, the optimal hp-adapted mesh for the accuracy the user chooses.

A commercial offshoot of COMCO's SBIR work is ProPHLEX Developers Toolkit™, introduced in 1995. The primary targeted users are major industrial corporations (for mechanical design applications) and universities (for research and mechanical engineering training).

Ford Motor Company has supported ongoing development to produce a companion product, PHLEXsolid™, and make it compatible with NASTRAN, the NASA-developed standard finite element analysis code used by Ford in auto design. Ford seeks to couple NASTRAN capabilities with the hp-adaptive and Residual Error Estimation features of the PHLEX library. Says a Ford official: "hp-adaptive finite element analysis is one of the key enabling technologies in Ford's drive to reduce the concept-to-market cycle."

Another spinoff product is UNISIM, a PHLEX-based code for analyzing underground flows in oil reservoirs. Developed by COMCO with funding from Texaco Inc, UNISIM is a finite element simulator that combines the hp-adaptivity feature with parallel processing. Designed to obtain accurate numerical pressure and saturation solutions associated with large oil fields, the software models the complicated dynamics of horizontal and vertical wells used to inject water into reservoirs and force more of the oil to the producing wells.

Ceramics Analysis

In the continuing quest for improved performance of aerospace systems, designers are exploring alternative materials that are stronger yet lighter than metals and have superior resistance to high temperatures. Among leading candidates are ceramics, which have already found service in aerospace propulsion systems in such high temperature Earth applications as diesel/turbine engine components and heat exchangers. They also offer performance benefits in products subjected to a lot of stress or wear, such as nozzles, valves, cutting tools, grinding wheels, bearings, and even artificial knee and hip joints.

Ceramic properties have improved a great deal in recent years due to advances in processing and composition, but they are inherently brittle and that leads to low strain tolerance and low fracture toughness. In addition, ceramic parts vary greatly in strength behavior; apparently identical parts can have vastly different fracture strength due to the variable severity of flaws in the materials. From the beginning of ceramics research, it was apparent that there was a need for a way of determining—in a product's design phase—the effects of stress and fatigue on a part, identifying all potential failure modes, and predicting lifetime capability. This was a challenge of first magnitude.

After more than 15 years of effort, Lewis Research Center provided a solution, a software package known as CARES/LIFE that accurately predicts the performance of brittle structures over time. The importance of this development is underlined by the multiple awards presented to the Lewis Research Center development team: a 1995 R&D 100 Award from *R&D Magazine* as one of the most significant technological advances of the year; a NASA Software of the Year Award; and the Federal Laboratories Consortium Award for Excellence in Technology Transfer. More than 300 organizations worldwide have used the CARES program in design of such widely diverse products as turbojet engines, valves for auto and truck engines, computer chips, cathode ray tubes and glass panels for office buildings. Some 65 companies are using the CARES/LIFE versions of the code.

A CARES/LIFE plot shows stress levels (red/yellow highest, blue/green lowest) on a turbine blade.

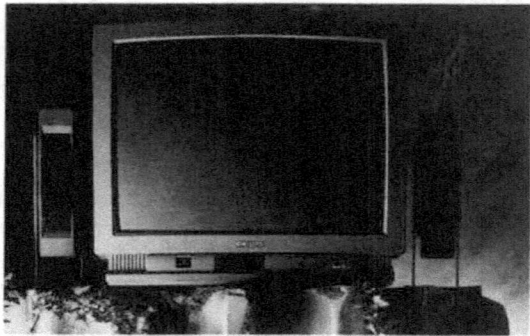

Philips Display Components Company used CARES/LIFE to analyze pressure forces on television tubes.

Lewis Research Center began work on the program in the late 1970s to serve the ceramics development needs of NASA, the Department of Energy (DoE), other government agencies and industry. At that time, no code for predicting ceramic durability existed. John Gyekenyesi, head of Lewis' Structural Integrity Branch, was assigned to devise one. Members of the team who shared the Software of the Year Award with Gyekenyesi include Lesley A. Janosik, Noel N. Nemeth, Lynn M. Powers and Jonathan Salem.

The CARES (Ceramic Analysis and Reliability Evaluation of Structures) program they developed offers a cost effective means for simulating tests of ceramic components; it enables a designer to test a variety of configurations for probability of failure, and to adjust the structure's geometry to minimize the predicted failure or maximize durability. The newest version of the software—CARES/LIFE, released in 1995—has the added capability of lifetime prediction.

Among the examples of the code's utility,

• Solar Turbines Incorporated, San Diego, California is using CARES/LIFE in a DoE-sponsored development program that seeks significant performance gains by replacing metallic hot section parts in a gas-fired industrial turbine with ceramics; ceramics last longer, offer more resistance, reduce friction and weigh less.

• Philips Display Components Company, Ann Arbor, Michigan analyzed glass television tubes with CARES/LIFE to determine the forces placed on the outside of an evacuated tube and the possibility of tube implosion. The research analysis concluded that the possibility of tube implosion is nearly zero and that an implosion protection band reduced maximum principal stress by 15-20 percent and further reduced the probability of failure.

• The CARES software is in use at AlliedSignal Turbocharging and Truck Brake Systems, Torrance, California, where the company has developed automotive turbocharger wheels for a diesel engine, the first design and large-scale development of ceramic turbochargers in the U.S. AlliedSignal has delivered more than 1,700 units to Caterpillar Tractor Company, Peoria, Illinois for on-highway truck engines.

• Among a number of automotive engine applications, TRW Inc.'s Automotive Valve Division is using CARES software to design ceramic poppet valves and engine components for passenger cars. General Motors Corporation is similarly using CARES for automotive applications.

CARES and CARES/LIFE were developed in the public domain and thus are readily available to industrial firms. The codes are distributed through NASA's Computer Software Management and Information Center (COSMIC)® at the University of Georgia.

TRW Inc. used CARES to design ceramic poppet valves and engine components.

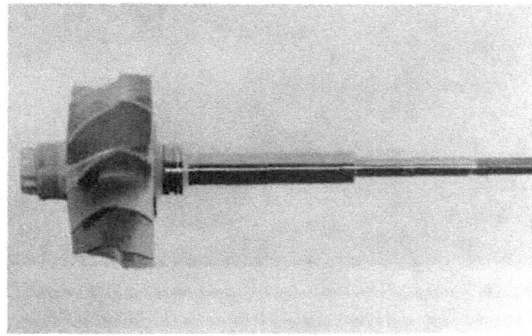

AlliedSignal's development of automotive ceramic turbocharger wheels was aided by CARES software.

Space Age Training

Among spinoffs in industrial productivity and manufacturing technology is an innovative system for more effective employee training

In the increasingly competitive global environment of the 1990s, a vital asset for any company is a skilled and efficient workforce. Entry level training and regular upgrading for experienced employees has become as important to a company's future as its end product.

Teledyne Brown Engineering (TBE), Huntsville, Alabama has introduced to the commercial market an innovative training aid designed to ease industry's instructional task. The company has developed a computer-based system for industrial process training that, says TBE, offers "significant improvements in effectiveness, standardization and quality, as well as cost reduction in comparison with the usual classroom and on-the-job training approaches." Known as the Interactive Multimedia Training (IMT) system, the commercial package is an offshoot of a system developed by TBE, under Marshall Space Flight Center management, to train astronauts and space operations personnel in the on-orbit operation of scientific experiments.

The experiments involved growing crystals as part of the USML-2 (U.S. Microgravity Laboratory) mission on Space Shuttle flight STS-73, which was launched on October 20, 1995. Carried within a pressurized Spacelab module that enables astronauts to work in a shirtsleeve environment, USML-2 embraced a wide-ranging array of crew-tended investigations of the behavior of crystals, fluids and other

Kathie Barnett of PPG Industries undergoes on-the-job training with the help of a new Interactive Multimedia Training system. Developed by Teledyne Brown Engineering, the system is a commercial derivative of a training system used in Space Shuttle science operations.

materials in the near-zero-gravity conditions of low Earth orbit.

The crew used a Teledyne Brown advanced, high temperature furnace for growing electro-optical and protein crystals. Space-grown crystals are larger and more uniform than crystals grown in Earth's gravity, and observation of these crystals during orbital flight offers scientists an opportunity to gain a better understanding of the crystals' molecular structures, research that can pave the way to a broad variety of practical applications. TBE's contract with Marshall Space Flight Center called for training the crew of STS-73, along with ground-based space operations technicians and investigators of crystal growth processes, in the complexities of the furnace's hardware and software and the scientific/operational procedures involved in crystal growth research.

Spacelab training normally requires sending trainees to special facilities where they use manuals, briefings and simulation to learn how to operate Space Shuttle in-flight experiments. It takes two years to prepare and conduct a training program, and it requires the participation of trainees and instructors at widespread facilities around the world.

Looking at rising trainee travel costs at a time of lower space funding, together with accelerated Shuttle schedules and reduced training time, NASA and TBE decided they had to find a better way to train space operations teams.

They found the better way in IMT, a system whose development is based on TBE's proven training philosophy that "people learn faster and retain more by participating actively in the learning experience." IMT presents information in a way that engages all the senses, using text, video, animation, voice, sounds and music. It allows learners to direct their own learning, at their own pace; they can repeat portions of the training program as many times as necessary to achieve complete understanding. Advanced types of simulations put learners in role-playing scenarios where they must react or interact as they would on the job; software driven simulations provide hands-on training, enabling learners to practice in a controlled environment with realistic feedback.

Among the advantages of multimedia training cited by TBE are its capability for simple explanation of difficult concepts; increased trainee comprehension and shorter learning time; standardized course quality and content, not dependent on the knowledge and skills of the individual instructor; and reduced or eliminated travel time and expense.

The USML-2 flight served as a test bed for evaluating the IMT approach; it turned out to be a highly successful "proof of concept" that established the value of multimedia training as an effective instructional tool. It prompted TBE to bring the system to the commercial market and offer industry a packaged course, customized to a particular industrial activity, for qualifying new hires and recertifying experienced workers.

The initial customer for the commercial course was another major aerospace firm, also located in Huntsville—PPG Industries Aircraft Products, the world's largest supplier of aircraft transparencies. TBE customized a program to meet the needs of PPG's high tech, job shop, production environment; it provides an overview that explains how the new worker's job fits into the company's overall production, emphasizes proper use of safety equipment, and describes a variety of problems that might be encountered in manufacturing PPG products and the proper way of correcting them. In addition, the software provides automated testing and feedback for operator certification.

PPG found that the interactive multimedia approach is a more effective training medium than prior methods, that new hires benefit substantially from the opportunity to study at their own pace, and that the program is easily modified to accommodate new equipment or improved production processes.

Lightweight Jack

Richard Dudgeon, Inc. (RDI), Bridgeport, Connecticut supplies hydraulic jacking equipment, along with engineering and technical support, for heavy lifting projects. The company rents and sells standard jacks and also designs and manufactures custom jacks for specific applications. Among the company's products are a line of high pressure jacks/load cells that evolved from a prototype system developed for a NASA heavy lifting project.

three giant antennas, one at each location, that support deep space (interplanetary) missions; each measures 70 meters in diameter, is 22 stories high and weighs some 8,000 tons.

In 1990, JPL discovered that modifications to the antennas over the years had redistributed weight to the point where bearing failure—and possible collapse of the antenna—was threatened. JPL asked RDI to devise a solution for simultaneous measurement of several key load points within the bearing housings of the 70-meter antennas, an initial step in correcting the alignment.

The job called for a heavy lifting load cell system of extraordinary capability. It had to be ultralight for portability (it had to be carried to the bearing housing level 120 feet above ground via narrow ladders); it had to be extremely powerful to lift the weight of the structure above the bearings; and it had to be thin enough to fit between load points that measure as little as .025 inches. Additionally, it had to be developed within a tight time frame.

RDI assembled a project team composed of its own experts augmented by those of a number of Connecticut firms that assisted in the design, development and ultimately the manufacture of a complete synchronous load cell system. Prototype development took only six weeks and the system was used successfully in the DSN project.

A group of Dudgeon 700-ton load cells/hydraulic jacks, each of which weighs only 79 pounds.

A comparison of the NASA technology load cell (red) and a 3,000-pound standard hydraulic jack of the same lifting capacity.

The project involved lifting segments of giant antennas in NASA's Deep Space Network (DSN), which is managed by Jet Propulsion Laboratory. The DSN is composed primarily of three communication complexes located at Goldstone, California; Madrid, Spain; and Canberra, Australia; they are about 120 degrees of longitude apart to permit continuous tracking of a spacecraft by at least one station at all times, despite Earth's rotation. Key facilities are

RDI subsequently commercialized the technology with its Dudgeon High Pressure Ultrathin Pancake Jacks/Hydraulic Load Cells, which are designed for applications requiring minimal lift but high portability. They are ultralight as well as ultrathin; a system capable of lifting 700 tons weighs only 79 pounds, where a conventional jack of similar lift capability might weigh half a ton. Available in a capacity range from 50 to 2,000 tons, they are intended for such applications as bridge weighing/lifting, heavy industrial and turbine weighing/positioning, and weighing/positioning of utilities and power plant equipment.

Magnetic Bearing

Magnetic bearings support moving machinery without physical contact, for example, they can levitate a rotating shaft and permit relative motion without friction or wear. Long considered a promising advancement, they are now moving beyond promise into actual service in such industrial applications as electric power generation, petroleum refining, machine tool operation and natural gas pipelines.

Among companies producing advanced magnetic bearing systems for industrial use is AVCON, Inc., Agoura Hills, California. AVCON offers a unique technological approach that evolved from the company's work on contracts with Lewis Research Center, and Marshall Space Flight Center. The technology developed in the NASA programs contributed to AVCON's ability to overcome the limitations of early magnetic bearing systems, namely large size and weight, high power consumption and cost. The company's product line embraces a family of very compact, lightweight, power efficient, low cost bearing systems.

Beginning in 1989, AVCON worked with Lewis to explore the possibilities of a magnetic bearing system for the turbopump of the Space Shuttle Main Engine. NASA decided to investigate magnetic bearings, which in theory—at that time—offered multiple advantages over conventional rolling element bearings.

AVCON worked initially with Lewis on development of a magnetic bearing system for a Cryogenic Magnetic Bearing Test Facility. The resulting AVCON development was extensively tested over a two year span and these tests provided a wealth of data on the performance of magnetic bearings under severe conditions. In this program, AVCON developed the basic hybrid magnetic bearing approach that characterizes its commercial products, an approach in which both permanent magnets and electromagnets are employed to suspend a shaft; the permanent magnets provide suspension, the electromagnets provide control. Analyses of AVCON bearing tests showed that a hybrid magnetic bearing was typically only one-third the weight, substantially smaller and dramatically less power-demanding than previous generations of magnetic bearings.

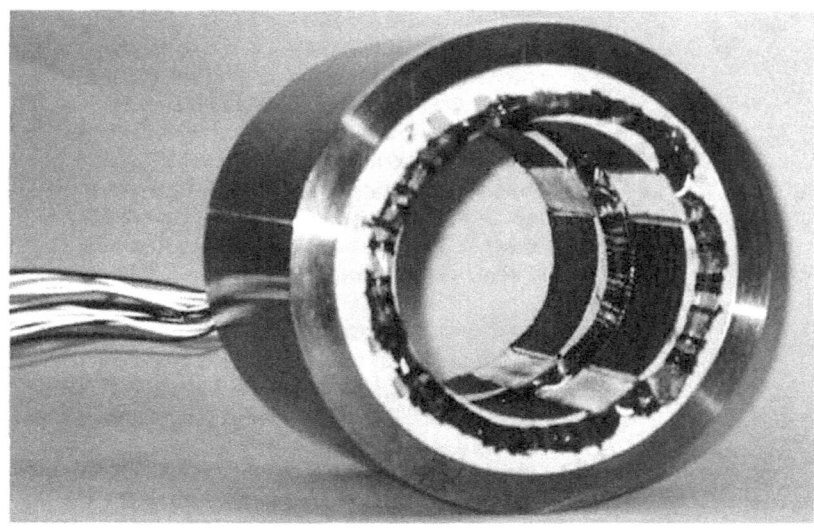

The AVCON magnetic bearing permits motion without friction or wear.

In 1993, Marshall Space Flight Center awarded AVCON a contract to fabricate a set of magnetic bearings, install them in a fixture representing a Space Shuttle Main Engine turbopump, and test them under simulated Shuttle mission conditions. Those tests are under way.

The technological base afforded by AVCON's NASA work, buttressed by additional R&D sponsored by the U.S. Air Force and a major turbine engine manufacturer, enabled AVCON to develop a unique "homopolar" approach to permanent magnet type bearings that, the company says, are significantly smaller than prior designs; additionally, their control electronics are a fraction of the weight of previous systems and power consumption is much lower than in all-electromagnetic designs. Among other advantages cited are virtually zero friction and therefore no lubricant requirement; no wear, no vibration; longer service life; and very high reliability because single point failure modes are eliminated.

The bearings were tested in a NASA turbopump unit.

Waterjet Impeller

North American Marine Jet Inc. (NAMJ), Benton, Arkansas manufactures waterjet propulsion systems for such craft as fishing boats, fire boats, houseboats and excursion boats. The company's newest line of marine jets incorporates NASA technology that, company president Leonard Hill expects, will make NAMJ significantly more competitive in a market long dominated by European and Pacific-area manufacturers.

Looking for ways to match the technological advances of his competitors, Hill attended a propulsion symposium at Marshall Space Flight Center (MSFC) and learned that technological assistance was available to him under NASA's Technology Transfer & Commercialization Program.

Leonard Hill of North American Marine Jet and Dinah Higgins of Marshall Space Flight Center display an impeller blade made by advanced rapid prototyping techniques.

Hill and his design staff sought advice from MSFC as to the efficacy of a proposed design for a new impeller, planned as the heart of a new line of marine jets. The Computational Fluid Dynamics (CFD) branch of MSFC's Structure and Dynamics Laboratory used advanced CFD techniques, including creation of a three-dimensional computer model of the impeller, to analyze the design and concluded that it would not provide the desired propulsive performance.

With Marshall input, NAMJ modified the design and a second analysis indicated that the redesigned impeller would meet or exceed NAMJ's specifications. MSFC then used a 3D computer model of the modified system to make a solid polycarbonate model of it. NAMJ was then able to have a metal prototype cast directly from a ceramic mold made directly from the polycarbonate.

The NAMJ experience not only exemplifies the kind of NASA assistance available to industry, it is also an example of the important time saving possible through adoption of MSFC's rapid prototyping technology. Under the center's Rapid Prototyping Program, which involves the direct production in three dimensions of a prototype from a computer-aided design, MSFC is developing technologies designed to enable sharp time/cost reductions in manufacturer's design-to-product development cycles. Ordinarily, NAMJ would have spent weeks creating a solid model of the impeller in a laborious, costly multistep process. The work of MSFC's Materials and Processes Laboratory and the use of rapid prototyping techniques allowed avoidance of many time-consuming and costly steps in creating the impeller model.

Heat Pipes

In the early days of space flight, NASA solved a major problem by teaming with Los Alamos Scientific Laboratory (LASL) in development of the heat pipe. The problem was that the Sun-facing surfaces of a non-rotating satellite became very hot while surfaces not exposed to the Sun became extremely cold, a temperature differential that threatened failure of electronic systems. The solution, used in virtually all spacecraft since its development, was the heat pipe, a tubular device in which a working fluid alternately evaporates and condenses, transferring heat from one region of the tube to another without external help.

This simple device offered a very broad range of practical Earth applications and NASA

Hot runner nozzles provide the consistent temperatures needed to make uniform parts.

prompted its broadest use by refining the technology and working with a number of other organizations on technology demonstrations. Heat pipe technology has become one of the most frequently tapped sources of spinoff applications. Some of the early users have further advanced the technology and broadened the range of applications through several generations of product development.

Among the latter is KONA Corporation, Gloucester, Massachusetts, a company formed in 1978 to manufacture hot runner systems for the plastics injection molding industry, using heat pipe technology that offered significant manufacturing and maintenance economies. KONA operated initially under a license from James M. Stewart, an independent consultant to the plastic industry, who had used the NASA/LASL technology as a departure point for his own development of patented "heat tubes."

KONA has continued to use the NASA technology as an integral part of its manufacturing equipment for such products as camera parts, kitchenware, auto components, TV cabinets and telephone parts. KONA has refined the technology through three generations of heat pipe advancement. The first was the KONA Nozzle, a heaterless injection nozzle designed to fit all injection molding systems; it was followed by a complete line of Hot Sprue Bushings. KONA also applies heat pipe technology to the company's Hot Runner Systems.

Temperature uniformity is critical in hot runner molding and the heat pipe, KONA says, is a way of getting it with multiple advantages over alternative systems. By offering a wide selection of hot runner nozzles and tips, KONA gives mold designers and moldmakers exceptional flexibility. KONA Hot Runner Systems are used throughout the plastics industry in the manufacture of products ranging in size from tiny medical devices to large single cavity auto bumpers and instrument panels. Sales of Hot Runner Systems account for 75 percent of all KONA sales.

Hot runner systems are used for production of a wide range of plastic products, from ice trays to medical devices.

Robot Tools

Robotics and Automation Corporation, Minneapolis, Minnesota manufactures equipment for robotic systems, in particular a variety of tools known as "end effectors," devices attached to the end of a robot arm for picking up, grasping, manipulating and transferring objects. The company traces its lineage to 1981, when a predecessor organization—INTEC Corporation—was formed to design and market robotic peripheral equipment. In May 1984 INTEC became Mecanotron Corporation and in 1989 Mecanotron was sold and incorporated into Robotics and Automation Corporation.

Among the company's newer products is the Automatic Robotics Tool-change System (ARTS), a system designed to meet growing demand for multiple task work cells for welding and plasma spray functions that require grinding and finishing; deburring, deflashing, routing, hole drilling or parts replacement; and multiple tool disk operations. The technology incorporated in ARTS systems was originally developed under contracts with Marshall Space Flight Center and with Rockwell International, one of NASA's principal contractors.

The ARTS systems were designed to work with the company's CFD (Constant/controlled Force Device) product line, a series of end effectors and bench mounted devices for controlling the constant pressure of abrasive tools used to deburr, grind, polish and finish products fabricated by welding, casting, molding, forging or machining.

Robotics and Automation Corporation's CFD line includes three end-of-arm devices and two

A Rocketdyne technician inspects welds on the Space Shuttle Main Engine created by an advanced robotic system.

bench-mounted devices. They do not require that the robot apply and control the force, only that it move along a normal programmed path over the work piece; the CFD applies and maintains the required processing pressure of the finishing media to the work piece.

When the surface to be finished is very rough and coarse, several different grades of finishing media may be needed, as well as different speed and power as the surface finish is transformed. To accommodate this multistep process within a single work cell, and with a single robot, Robotics and Automation Corporation developed the automated tool-change system.

The ARTS-I is being used in industrial systems with six tool positions ranging from coarse

sanding disks and abrasive wheels to cloth polishing wheels with motors of various horsepower. The ARTS-II allows a robot to change welding torches automatically, or to exchange a welding torch for a CFD end effector to finish a welded assembly with a welding robot; using a second tool-changer (ARTS-I) enables finishing the surface conditioning process.

Robotics and Automation Corporation has sold more than 90 robotic work cells using CFD/ARTS devices, about one fourth of them in the plastics industry. The largest single user category is fabricators of plastic body parts for the auto industry; other uses range from fabrication of radar domes by Texas Instruments to advanced composites at Aerospatiale in France.

The quick disconnect system allows changing tools with hydraulic, pneumatic or electric power.

The tool rack of the Automatic Robotics Tool-change System includes a two-finger gripper, a grinder, a coated abrasive brush and a welding torch.

Advanced Welding Torch

In the late 1970s, when the Space Shuttle was in early development, NASA recognized that the then-existing welding techniques were inadequate for the job of joining the huge aluminum segments of the Space Shuttle External Tank. Accordingly, Marshall Space Flight Center (MSFC) initiated development of Variable Polarity Plasma Arc (VPPA) welding. The VPPA concept dates to 1947, but it had never been fully developed; it employs a variable current waveform that enables the welding system to operate for preset time increments in either of two polarity modes for most effective joining of troublesome light alloys such as aluminum and magnesium.

In the course of VPPA development, it became apparent that the technique had broad potential for improving weld reliability and lowering costs not only in NASA work but in many private industry applications. Since there were no suitable commercially available tools for VPPA welding, MSFC expanded the development effort to include a technology transfer project designed to make VPPA available to the private sector.

A key part of this effort was development of a welding torch that would have dual utility, as a component of NASA's External Tank welding system and as a component of derivative systems for commercial applications. MSFC awarded the torch contract to B&B Precision Machine, Owens Cross Road, Alabama. Working in cooperation with MSFC's Materials and Processing Laboratory, B&B developed and patented a Shuttle-use torch that won a 1987 NASA Inventor of the Year Award for Bob Dempsey of B&B, Ernest Bayless and Sam Clark of MSFC.

The B&B Precision Machine Variable Polarity Plasma Arc welding torch.

MSFC and B&B continued development of VPPA. A major step in the late 1980s was a program to fully automate the system and eliminate the hand of the welder on the controls entirely. In 1989, a NASA decision to change the material of the External Tank triggered a new B&B development. The new alloy in some cases required "tack" welds prior to robotic seam welding; since tack welds are performed by hand, B&B was assigned development of a smaller version of the torch that would be easier to manipulate and would meet the needs of applications where access was limited. B&B delivered a prototype small torch in 1992.

The small torch, which has attracted considerable interest in the commercial sector, has the same features and advantages as the original torch, but it fits in approximately half the space. It is in commercial service with Whirlpool Corporation for sheet metal welding of major appliance parts, where the torch's production line dependability is a significant asset. Offering such advantages as multiple cost reductions and eventual reduction of requirements for x-ray inspection of welds, the microprocessor-controlled VPPA system is in use at the plants of such industrial giants as Babcock and Wilcox, Boeing, General Dynamics, Lockheed Martin and McDonnell Douglas. The system and the B&B torch continue to make all the welds in the External Tank and they have been selected as the preferred welding approach for the International Space Station.

A small version of the B&B torch is used in commercial sheet metal welding.

Protective Coatings

Ameron International Protective Coatings Group, Brea, California has been for more than half a century among the world's leaders in developing and producing protective coatings. The company has long been a supplier to NASA for coatings to protect space launch structures and other facilities. Beginning in 1990, Ameron developed a special formulation to meet NASA requirements for space launch pad coating. That technology is now being used in the company's commercial product line.

NASA wanted a coating that could withstand the 5,000-degree Fahrenheit temperatures generated by the blast of the Space Shuttle's rocket engines. The coating had to remain intact at that temperature for at least 10 minutes, and it had to insulate the launch pad so that its steel would not heat above 150 degrees Fahrenheit and buckle. Additionally, NASA wanted a sprayable coating that would cope for long periods with the heat, humidity and ultraviolet attack of the intense Florida sunlight.

For the NASA assignment, Ameron created an extra-high-temperature-resistant formulation of its Engineered Siloxane™ (PSX™) chemistry, which employs an inorganic silicon-oxygen structure that, Ameron states, is stronger and more reliable than the carbon-based structure in organic polymers. The formulation—known as Amercoat® 3335—won NASA approval and brought NASA a bonus: PSX products formulated from this technology can be applied in one coat directly over the inorganic zinc primer, with no need for a mid-coat, therefore it offers reduced application time and labor for general structural steel applications. Amercoat 3335 has been applied to the launch pads for the McDonnell Douglas Delta rocket at Cape Canaveral Air Force Base. Additional applications to other launch pads at the Kennedy Space Center/Cape Canaveral complex are planned.

The PSX technology that was tailored to NASA requirements is incorporated in some general purpose formulations for the company's commercial customers. One product, PSX-700, is designed for exceptional weatherability, corrosion control and long-lasting protection with only one application; it is intended for such uses as bridges and marine structures, industrial plants, tanks and piping, and on transportation vehicles, including boats and barges. PSX-738 is a product designed to withstand twice as much continuous heat as conventional heat-resistant coatings (more than 2,000 degrees Fahrenheit), and it is capable of protecting both carbon steel and stainless steel, even under insulation.

*Engineered Siloxane and PSX are trademarks of Ameron International Corporation.
*Amercoat is a registered trademark of Ameron International Corporation.

An Ameron formulation designed for protection of NASA launch pads is in service on the launch pads for Delta rockets.

Ameron's PSX 700 coating is an industrial-use adaptation of the launch pad coating.

Small Business Innovations

In 1982, Congress established the Small Business Innovation Research (SBIR) program as a means of increasing opportunities for small businesses to participate in federal R&D activities. A related objective was to stimulate conversion of government-funded R&D into commercial applications; that benefits the U.S. economy—in terms of jobs created and contribution to the Gross Domestic Product—when the SBIR project generates a commercial spinoff.

Each technology generating agency of the government sets aside a percentage of its R&D budge for SBIR projects. There are 11 such agencies, each administering its own program independently under policy guidelines set by the Small Business Administration.

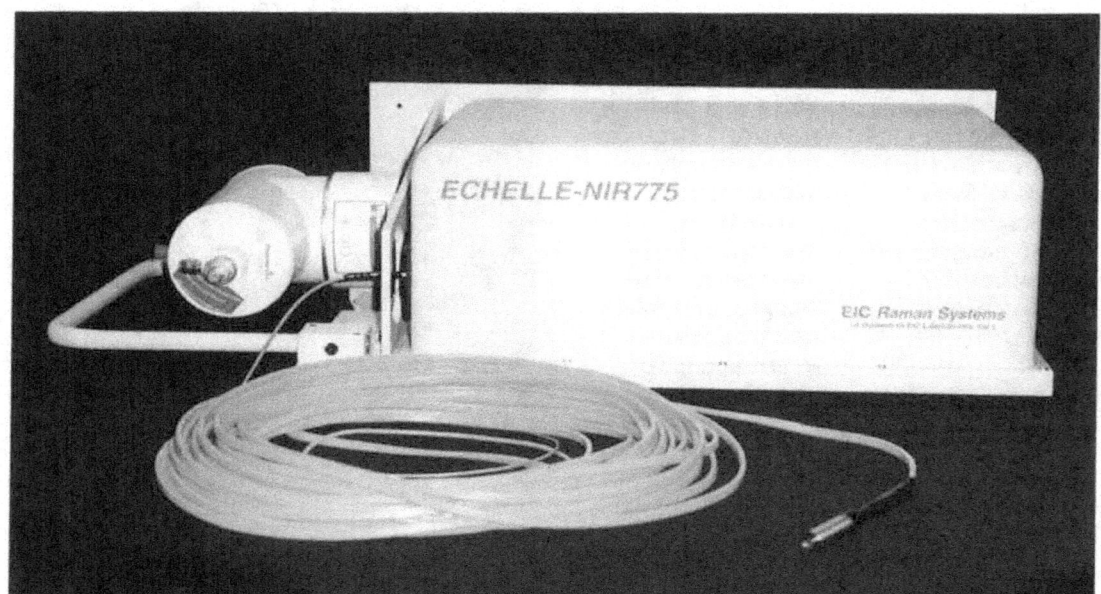

NASA's SBIR program has been eminently successful. It has provided the agency an additional source—beyond traditional aerospace firms—of R&D talent and innovative thought. Hundreds of new systems that advance NASA's capabilities for aerospace research and operations have emerged from the SBIR program. About one of every three SBIR projects results in a commercial spinoff.

EIC Fiber Optic Raman Spectrograph

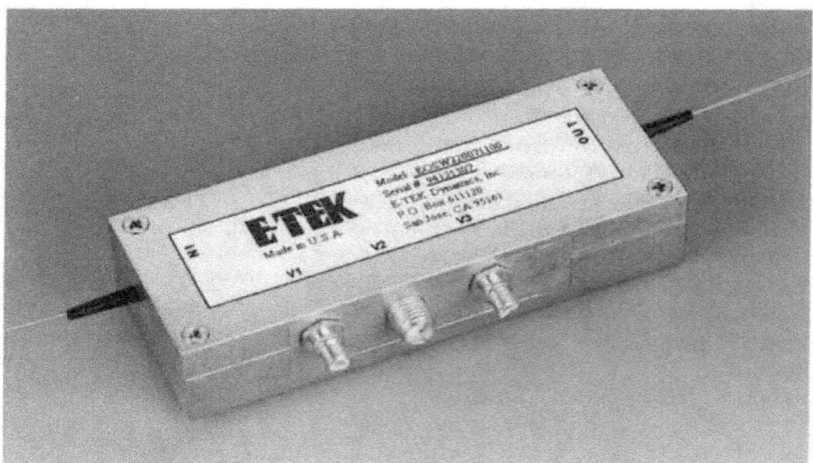

E-TEK High Extinction Ratio Electro-optic Switch

Among representative examples of SBIR projects in the field of industrial productivity is a family of spectroscopic instruments developed by EIC Laboratories, Inc., Norwood, Massachusetts. EIC's instruments are based on Raman spectroscopy, a laser-based measurement technique that provides—through a unique vibrational spectrum—a molecular "fingerprint." Raman offers an advantage over infrared absorption techniques in its ability to function in aqueous environments. EIC is combining optical fiber technology with Raman methods to develop sensors that can be operated at a considerable distance from the laser excitation source and the spectrographic analysis instrumentation.

That technology was substantially advanced under a NASA SBIR contract designed to produce a Raman spectrograph with fiber optic sampling for such space applications as sensing hazardous fuel vapors, monitoring hydrogen gas, and making on-board rapid analyses of chemicals and minerals.

EIC successfully developed the NASA system then, using its own capital, refined the technology to create a commercially available Fiber Optic Raman Spectrograph and an associated patented RamanProbe™, a fiber optic probe that can make measurements up to 500 meters distant from the spectrograph. The system has no moving parts and is 10 times more compact than prior equipment.

Among industrial applications of the system are process control, polymer processing, analyzing liquid mixtures, corrosion analyses, monitoring hazardous materials, quality assurance assessments, and use in the manufacture of pharmaceuticals and semiconductors. The Raman spectrograph and probe system has been a singularly successful technology transfer, one that brought EIC $3 million in sales over a two-year period and resulted in creation of a new company division—EIC Raman Systems—to provide commercial Raman instruments and services.

Another example of a successfully commercialized NASA SBIR project is the work of E-TEK Dynamics, Inc., San Jose, California. Founded in 1983, E-TEK initially confined its activities to

R&D in the fiber optic, microwave/millennium wave, and integrated opto-electronics disciplines. Since 1990, however, the company has been manufacturing and marketing a line of integrated optic and fiber optic components and instruments; annual sales grew from $2 million to $32 million in the 1990s and employment increased about tenfold.

E-TEK has worked on a number of NASA SBIRs. One of them, sponsored by Kennedy Space Center (KSC), called for a new line of electro-optic switches for fiber optic communications and optical signal processing applications. KSC wanted faster switch speeds, substantially smaller switches and much improved temperature stability.

Under the SBIR, E-TEK developed a line of microfabricated switches with nanosecond speed, extremely low crosstalk and the requisite temperature stability. The technology developed in the KSC project was integrated in E-TEK's commercial product line, specifically in a High Extinction Ratio Electro-optic Switch for high speed communications and optical signal processing, a 4x4 Switch Module, and a Programmable Fiberoptic Switch for routing optical signals, automatic optical testing and fiber optic communications.

(Continued)

*RamanProbe is a trademark of EIC Laboratories, Inc.

E-TEK Programmable Fiberoptic Switch

Small Business Innovations *Continued*

Before each Space Shuttle flight, an extraordinary amount of ground processing is required to assure safe and effective operation of the Shuttle and its payloads. To shave the time, lower the costs and increase the efficiency of these ground processing operations, NASA has turned increasingly to automated systems. But robotic systems pose problems, too; there is the chance that an errant robot might damage critical flight hardware.

Kennedy Space Center (KSC) saw a need for an obstacle detection system to insure that robots avoid collisions in their workplace. Accordingly, KSC sponsored a Small Business Innovation Research (SBIR) program with Merritt Systems, Inc. (MSI), Merritt Island, Florida, a research and consulting firm with special expertise in whole-arm proximity sensing technology for dextrous robot manipulators. From 1991 to 1995, KSC and MSI worked together on four SBIR projects to develop sensing and control technology for whole-arm robot manipulators.

From this work, Merritt Systems has developed a unique sensing architecture that allows retrofitting existing types of robots with a whole-arm obstacle detection and avoidance system that offers applications for NASA, other government agencies and their contractors, and also has broad potential for commercial use in such areas as robotic manufacturing systems, remote hazardous waste cleanup, and high value robotic tasks in constrained environments.

The key elements of the MSI system are the innovative sensorSkin and its associated compact, low cost smartSensor modules. Made

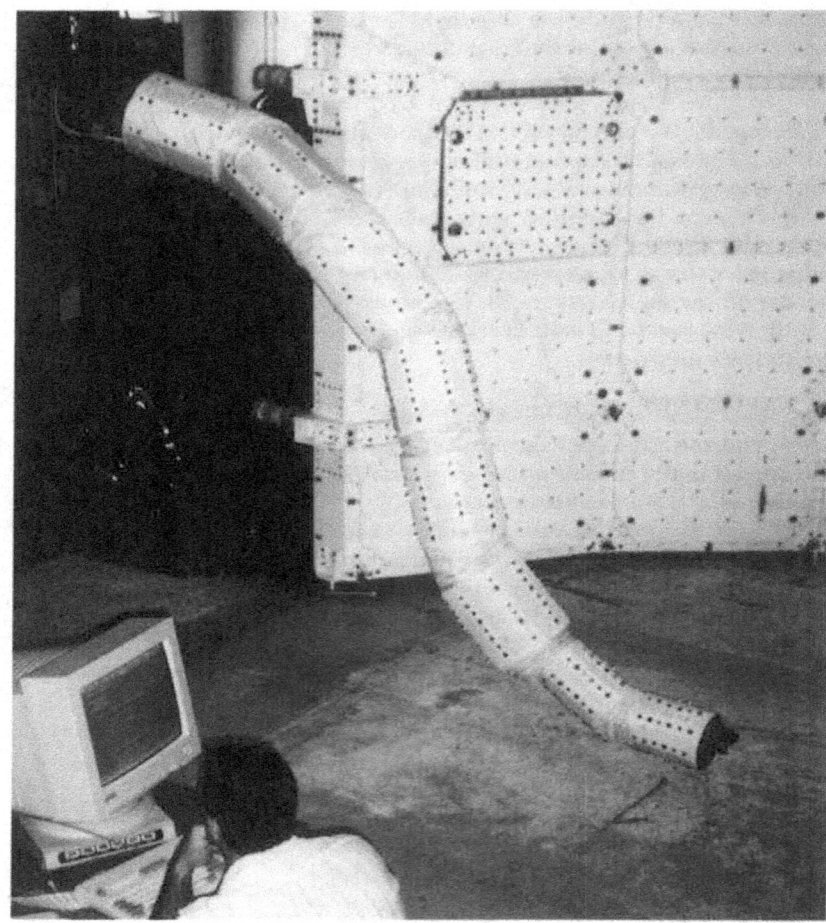

Merritt Systems' Tom Pigoski monitors the MSI Obstacle Detection System as it directs a robot arm away from an obstacle.

of flexible material, the sensorSkin can be cut and shaped to fit the mechanical arm of a robot. Within the skin is a distributed array of sensors—more than 1,000 of them, including proximity, motion, contact/force and temperature sensors—networked interchangeably over a single four-conductor wire. Each compact module contains on-board intelligence to perform all the analog processing and handle all communications between the module and the control computer. An on-board microcontroller processes the sensor information and transmits only relevant information back to the control computer.

MSI also developed a control algorithm for active collision avoidance during robot arm motions; as obstacles are encountered, a manipulator arm will react to avoid the obstacle while at the same time maintaining its desired end-effector position and orientation. The algorithm is incorporated in MSI's Robot Simulation and Control Environment (RSCE), which also provides kinematic control and three-dimensional graphic animation of robotic devices. The RSCE is designed as a tool for robotics training, analysis, design and control applications.

Robots incorporating the MSI sensorSkin and smartSensor technology are now being used in Space Shuttle processing. MSI proceeded to commercialize the technology; the initial commercial systems were delivered in 1995. MSI is investigating new applications in a variety of fields, including industrial process monitoring, building security, medical instrumentation and intelligent automation applications.

(Continued)

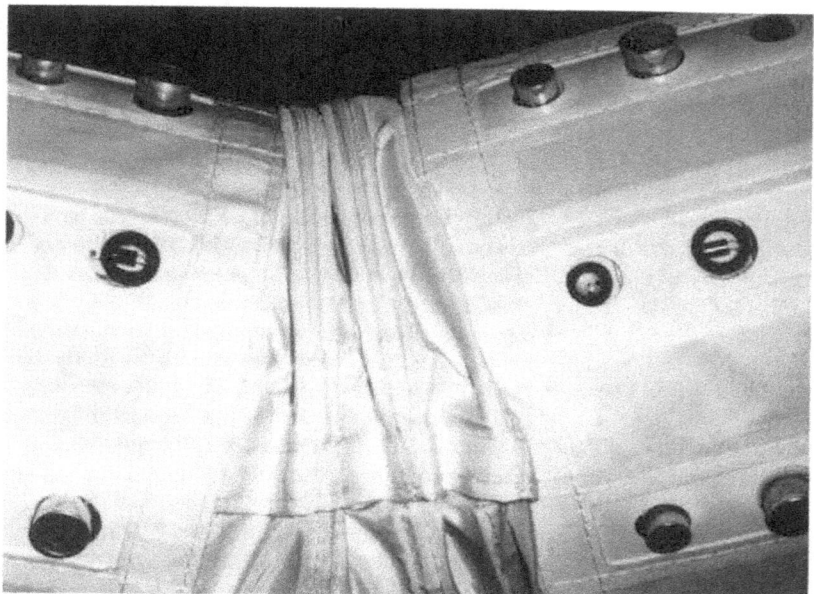

A closeup of the sensorSkin, which can contain more than 1,000 compact smartSensors providing intelligence to a control computer.

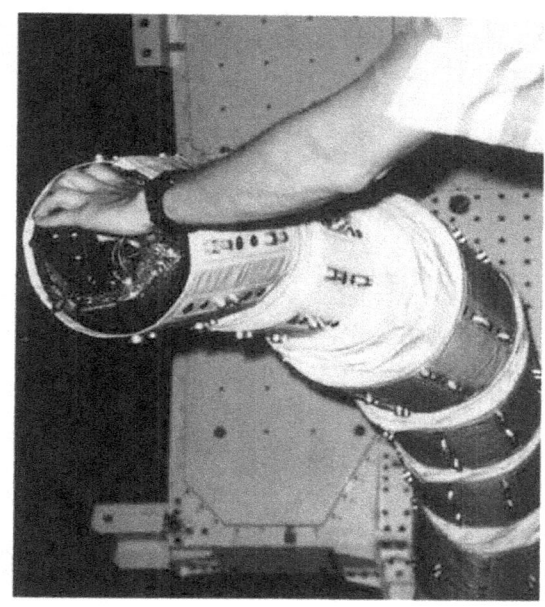

An operator adjusts the Obstacle Detection System in a robot arm.

Small Business Innovations *Continued*

Metal hydrides are chemical compounds formed by the reaction of hydrogen with metals, alloys or intermetallic compounds. Metal hydrides that react at room temperature were discovered in the 1960s. Even before that they had found many practical applications, for example, in processing steel, coating and bonding processes, in the preparation of metal powders, and in portable hydrogen generators for weather balloons.

Hydrogen Consultants, Inc. developed a compact metal hydride container for extended storage of industrial-use hydrogen.

Among the properties of metal hydrides is their ability to store hydrogen in a solid state, an area that has not been widely exploited. Since hydrogen is a common spacecraft propellant, but difficult to store, NASA was interested in the potential of metal hydrides as a means of storing hydrogen in solid state and thus avoiding the hazards of compressed gas or the complexity and boiloff of liquid hydrogen.

Marshall Space Flight Center (MSFC) awarded a Small Business Innovation Research (SBIR) contract to Hydrogen Consultants, Inc. (HCI), Littleton, Colorado to explore the utility of metal hydrides in spacecraft hydrogen systems. A follow-on Phase II SBIR directed HCI to design and develop two prototype hydride systems identified as promising in the Phase I effort: a Long Term Hydrogen Storage System for space use and a Metal Hydride Refrigerator for possible use aboard the International Space Station.

HCI delivered prototype systems to MSFC for testing. The Metal Hydride Refrigerator is thermally powered and can operate off a low to moderate source of waste heat, which makes it ideal for spacecraft applications where electric power carries a big weight penalty. Clearly, the refrigerator also has broad potential for Earth applications in view of the fact that it requires no compressor, a significant advantage in light of the planned phaseout of terrestrial freon-based refrigeration systems.

The Long Term Hydrogen Storage System delivered to MSFC enables storage of 10 pounds of hydrogen in a vessel only 15 inches in diameter and 32 inches long; its principal advantages are extended storage time and its compactness, compared with conventional cryogenic (very low temperature) storage. HCI has drawn upon its SBIR work to produce a commercial derivative of the technology under the trade name SOLID-H.

SOLID-H systems are a series of compact containers in which hydrogen is stored in solid state, offering an attractive alternative to large high-pressure cylinders and small disposable cylinders in industrial storage applications. Hydrogen gas is converted to solid state by a chemical absorption process in which the gas reacts with powdered metal crystals within the container to form metal hydrides. The hydrogen can be stored at room temperature and released without high pressures by decomposition of the metal hydrides, which liberates the hydrogen while returning the crystals to their original state. Among SOLID-H advantages cited by HCI are economy, safety, rechargeability of the containers, and their compactness; the containers are 9 ½ inches high, 6 to 6 ½ inches wide at the base, weigh only 4 ½ to 9 pounds, and have capacities from 40 to 140 liters.

Another example of a NASA SBIR project that spawned commercial products is the work of Barr Associates, Inc., Westford, Massachusetts. Established in 1971, Barr has been a supplier of optical filters to NASA, the European Space Agency, and other space-oriented organizations since the company's inception. Barr has provided filters for instruments used in such NASA projects as the Hubble Space Telescope, the Galileo spacecraft, the Cassini planetary explorer to be launched in 1997, and the Multiple-angle Imaging Spectroradiometer slated for space service in 1998. The company's filters have flown on, or are scheduled to fly on, more than 40 space-based instruments.

In 1989, Barr was awarded an SBIR contract by Jet Propulsion Laboratory (JPL) to develop and fabricate advanced technology, image quality, space-qualified ultraviolet (UV) interference filters. UV filters are thin film-coated windows that act like sunglasses on instruments to enhance scientific observations, such as ozone studies, planetary atmospheric compositions, or the chemical reactions of environmental pollutants. Over a two-year period, Barr developed an advanced ion-assisted deposition process, which enabled creation of filters that eliminated certain technical problems associated with earlier filters and also broadened the range of environments in which the filters can be used.

Barr has since refined the JPL deposition technology and used it in a commercial line of filters that have utility in such applications as fiber optic communications, hand-held spectrometers, tabletop laboratory analyzers and a variety of industrial applications. The filters are stable, durable, provide high spectral performance, and can be fabricated in miniature sizes for portable instruments.

A Barr Associates technician unloads a filter from a chamber where thin film coatings are applied to ultraviolet filters.

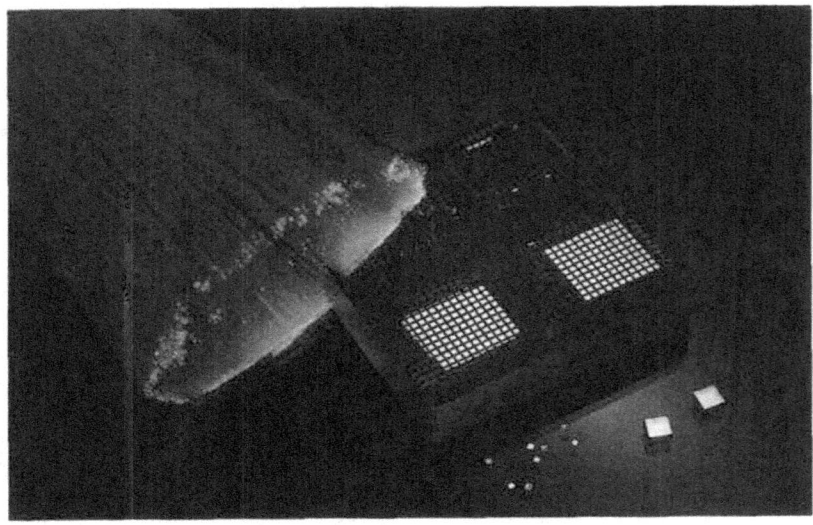

Barr developed an advanced ion-assisted physical vapor deposition process to produce high performance filters for fiber optic communications and other applications.

Technology Transfer & Commercialization

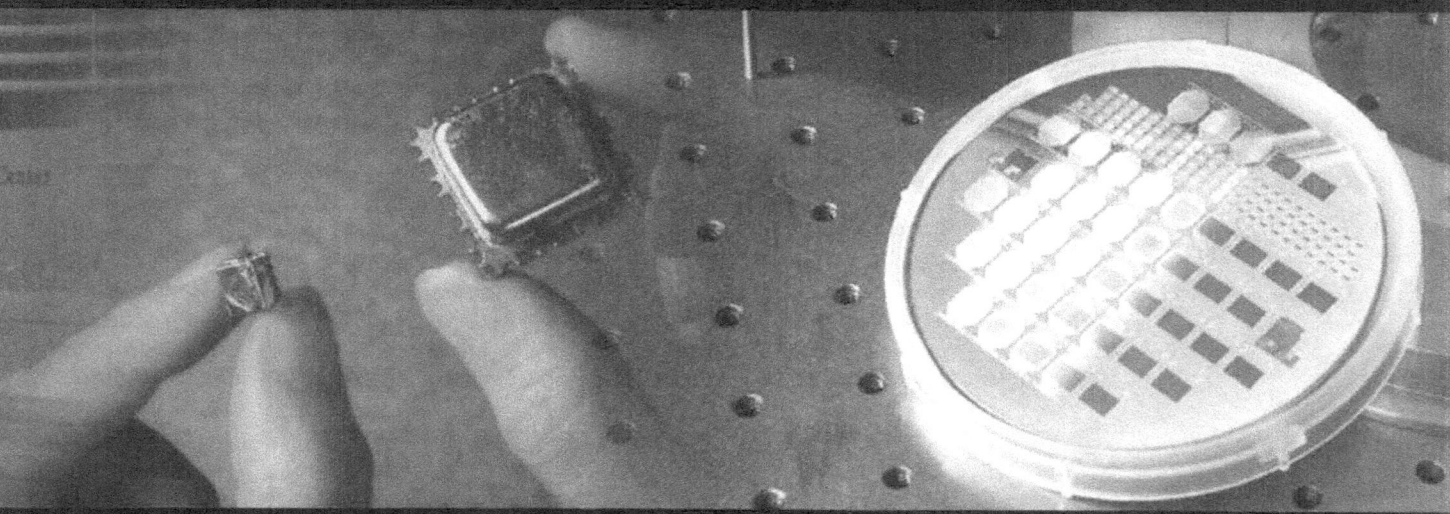

A description of the mechanisms employed to encourage and facilitate practical application of new technologies developed in the course of NASA activities

Putting Technology to Work

A nationwide technology transfer network seeks to broaden and accelerate secondary use of NASA-developed technology

Because they are challenging and technologically demanding, NASA programs generate a great wealth of advanced technology. This bank of technology is a national asset that can be reused to develop new products and processes, to the benefit of the U.S. economy in new companies, new jobs, and the resulting contribution to the Gross Domestic Product.

Such "spinoff" applications do not happen automatically. It takes a well-organized effort to put the technology to work in new ways and to reap thereby a dividend on the national investment in aerospace research.

NASA accomplishes that end by means of its Technology Transfer & Commercialization Program, which employs a variety of mech-

anisms to stimulate the transfer of aerospace technology to other sectors of the economy. The program is managed by the Commercial Development and Technology Transfer Division of NASA's Office of Space Access and Technology. Headquartered in Washington, D.C., the division coordinates the activities of technology transfer organizations located throughout the United States.

Among the most important mechanisms are the technology transfer and commercialization offices at NASA's 10 field centers. These offices differ somewhat from center to center, but generally their jobs involve promoting transfer and commercialization of technology that has significant potential for secondary use. Repre-

Nona Minnifield, Commercial Research Manager in Goddard Space Flight Center's Office of Commercial Programs, displays samples of innovative, Goddard-developed camera technology that has potential for commercial applications. The gold-rimmed device in foreground is a CCD (Charge Coupled Device) Camera that takes electronic pictures with high resolution; among other applications, it offers special promise in advanced mammography. The camera, its sensors and a wafer of sensor chips are shown in close-up at right.

sentative of this type of activity is the work of the Office of Commercial Programs' Technology Transfer & Commercialization Office at Goddard Space Flight Center (GSFC).

Located in Greenbelt, Maryland, GSFC is a facility with personnel expertise in all phases of space operations, including the design, construction and test of spacecraft; the ability to operate, track and communicate with satellites in orbit; and a capability for analyzing and disseminating satellite data. Operational since 1959, GSFC has managerial responsibility for the NASA Tracking and Data Relay Satellite System ground stations at White Sands, New Mexico; the Goddard Institute for Space Studies in New York City; and NASA's Wallops (Virginia) Flight Facility, a special base for

suborbital research with sounding rockets. Additionally, the Space Telescope Science Institute in Baltimore, Maryland, is operated under contract to GSFC.

While building a reputation as one of the world's foremost space research organizations over more than three decades, GSFC has also compiled an outstanding record for transferring NASA-developed technology to industry and other potential users. The center boasts a lengthy list of successful transfers, among them LIDAR technology, widely used in the remote sensing industry; microelectronics technology, which has found many applications, particularly in biomedical equipment; compliant cable technology, which has benefited the health care and utilities industries; and a wealth of innovative software employed in many industries, in particular the automotive industry.

Continuing this tradition, Goddard's Technology Transfer & Commercialization Office serves as a point of liaison between GSFC and those organizations or individuals who might be able to make advantageous use of NASA technology. The mission of the Technology Transfer Office is twofold: to develop and implement effective marketing strategies and to facilitate partnering relationships with U.S. industry, other government agencies and academia for transfer/commercialization of GSFC-developed technologies.

The following pages contain a summary of the mechanisms employed by NASA to promote technology transfer and commercialization, including the structure of the nationwide network, the types of assistance provided by the Regional Technology Transfer Centers, and the technology transfer activities of NASA's field centers; the summary focuses on activities within GSFC's Office of Commercial Programs and the Technology Transfer Office as representative of today's technology transfer work performed by the field centers and other groups within the NASA network.

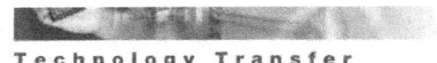
Technology Transfer & Commercialization Activities

An important element among NASA mechanisms for promoting technology transfer is the Technology Transfer Office, or TTO. TTOs are technology transfer experts based at each of NASA's 10 field centers and one specialized facility who serve as regional managers of the Technology Transfer & Commercialization Program.

At Goddard Space Flight Center (GSFC), the Technology Transfer & Commercialization Office is part of a larger office known as the Office of Commercial Programs (OCP). This office embraces three branches, one dedicated to coordinating the center's Small Business Innovation Research program, another concerned with the center's activities relative to commercial use of space, and the Technology Transfer & Commercialization Office. Dr. George Alcorn heads the broader OCP and is the center's chief TTO; Nona Minnifield, formally titled Commercial Research Manager, heads the Technology Transfer Office. They are backed by commercial technology engineers Evette Conwell and Carmon Parkinson; patent counsel Dennis Marchant; patent attorney Keith Dixon; legal technician Tammie Howcott; and secretary Michelle Wallace.

The duties of the TTOs vary somewhat among NASA centers and other units of the agency, but the organization and activities at GSFC are representative of the "new way of doing business" in recent years, in which NASA is

The chief of Goddard's Office of Commercial Programs, Dr. George Alcorn, oversees the center's technology transfer efforts, such as the Earth Alert project.

broadening the scope of its technology transfer efforts and placing greater emphasis than in prior years on commercialization of NASA technologies. The TTO's main responsibility is to stay abreast of research and development activities at his center that have significant potential for generating transferrable knowledge. He assures that the center's professional people identify, document and report new technology developed in the center's laboratories and, together with other center personnel, he monitors the center's contracts to see that NASA contractors similarly document and report new technology, as required by law. This technology, whether developed in-house or by contractors, becomes part of the NASA bank of technology available for transfer.

Technology transfer activities at GSFC embrace three areas of effort: inreach, outreach and marketing. The Inreach Program is key to heightening the awareness and stimulating the participation of center personnel in technology transfer and commercialization. It also involves center-wide technology training courses; ongoing consultation with GSFC scientific and technical personnel; annual new technology recognition programs; center-sponsored colloquia; and directorate-sponsored employee development programs.

The Outreach Program is structured to identify and increase awareness of GSFC technological resources to businesses, universities and other public entities within the U.S. This is accomplished through professional and trade conferences and meetings; on-site and regional commercialization workshops; and technology demonstrations, facility tours and consultation with industry groups. The TTO also works to communicate GSFC's resources within the state and local government complex, through involvement with economic development offices, chambers of commerce, high tech councils, educational institutions, and scientific and technical organizations.

In the marketing function, the Technology Transfer & Commercialization Office works with an internal and external network to disseminate GSFC technology to the widest possible audience. This effort, framed around the Inreach and Outreach programs, employs electronic and print media to communicate the development and availability of new technologies with potential commercial applications. Such media include database systems, the Internet, videos, advertisements and brochures. GSFC's Technology Transfer & Commercialization Office also publishes information on its technological developments in several professional and trade publications.

One mechanism for announcing developments with commercial potential is the "Technology Opportunity Announcement," usually a brief flyer that describes a technology originally developed for NASA use by the center but available for transfer. A recent example: GSFC developed a Holographic Optical Element (HOE), a receiver telescope and scanning mirror to measure clouds and aerosols from a ground-based platform. GSFC's Technology Opportunity announcement described the technology, listed a number of potential commercial applications (laser probes/scanners, optical communications, robot vision systems, promotional attractions) and invited interested firms to submit proposals for transfer and commercialization of the technology.

The GSFC Technology Transfer & Commercialization Office also initiates partnerships with industry, other government agencies, or academic institutions to apply NASA-developed technologies to industrial and national needs. One aspect of partnering is facilities utilization, where specific GSFC facilities are made available for use, on a non-interference basis, by any of the partnering entities for R&D, analyses, and performance testing of products. The Technology Transfer & Commercialization Office coordinates and participates in both industry-led and government-led cost-sharing development programs.

Technology Applications

Engineering applications projects are efforts to create beneficial new products, or to solve significant public sector or industrial problems, through redesign or reengineering of existing technology. Some projects stem from requests for assistance from government agencies or industrial firms, others are generated by NASA technologists who perceive possible problem solutions or useful product developments by adapting NASA technology to a particular need. NASA also employs an applications team composed of scientists and engineers representing different areas of expertise, who identify problems, submit them to NASA centers for review, then assist the centers in adapting the technology. Engineering applications projects are generally conducted in cooperation with a partnering organization on a cost-sharing basis.

The camera pictured on page 116 exemplifies an application project undertaken by Goddard Space Flight Center (GSFC). The gold-rimmed device is a 4K × 4K Mosaic CCD Camera, used by NASA for ground-based astronomy at Kitt Peak National Observatory, Tucson, Arizona. The engineering applications project, in which GSFC is teamed with the University of Massachusetts, involves adapting the technology for use in mammography (cancer-screening breast examination).

Mammography is normally accomplished by x-ray examination of the internal structure of the breast. The 4K × 4K device offers significant improvement over the x-ray technique because it is an electronic camera that takes pictures without film by converting light directly into

This is a special version of the Earth Alert System disaster warning receiver designed for use in Hawaii; the emergency symbols are the same as those used in the Honolulu phone book. The system was developed as a partnership applications project of Goddard Space Flight Center and Scientific and Commercial Systems Corporation.

Satellite-relayed warnings are picked up by this solar-powered mobile tower/trailer antenna system and broadcast to the handheld personal receivers. The dish antenna is the satellite downlink antenna; the warning transmission antenna is on the 37-foot mast.

digital images; digitizing the image enables real-time display on a screen and allows technicians to zoom in on a specific feature and to enhance or otherwise manipulate the image to extract additional information. Continuing work on this application is aimed toward development of a display compatible with existing x-ray systems for clinical use.

Another example of an engineering applications project, one that also offers life-saving potential, is a disaster warning device for isolated populations—such as the inhabitants of islands in the Pacific—where there are insufficient means of delivering warnings of severe weather or other natural disasters. This project draws on NASA weather satellite technology to provide a fast, effective personal warning system for individuals threatened by a disaster, such as a hurricane, tornado, forest fire or flood. In this development, GSFC is teamed with Scientific and Commercial Systems Corporation (SCSC), Beltsville, Maryland. SCSC is a Goddard support contractor producing electronic schematics, circuit boards and integrated systems for such applications as steerable antenna controls and satellite downlink data processors.

Called Earth Alert, the system provides direct transmission of severe weather warnings to a handheld device that uses symbols, maps or text messages to indicate the nature and status of the emergency and allow the users to take survival actions. The system consists of a central trailer-mounted transmitter plus an appropriate number of handheld receivers, distributed to individuals, homes and offices in the isolated area. When a disaster threatens, an initial warning is sent to a National Oceanographic and Atmospheric Administration weather satellite in geostationary orbit. The satellite rebroadcasts the message back to Earth, where it is picked up by an Earth Alert ground station and broadcast, via the central transmitter, to the personal receivers in the area. Each Earth Alert system is tailored to a specific locale. The handheld beeper-like device, the mobile central transmitter, and the software/hardware for decoding and display were developed by SCSC.

Goddard and SCSC have modeled an Earth Alert System for the Republic of Fiji and developed a special variant of the receiver for use in the islands of Hawaii. The receivers were modified to report four threats—hurricane, tsunami, flash flood and high surf—specified by the State of Hawaii Office of Civil Defense. Tests demonstrated communications compatibility with standard radio and satellite systems. SCSC is continuing its partnership with GSFC by working on a design for a standardized VHF receiver. The company is offering commercial Earth Alert Systems tailored to specific locales, including the central transmitter, documentation, training and a block of 1,000 receivers.

A closeup view of the Earth Alert System mast showing the electronics container (box in center of photo) and the warning transmission antenna (top of mast).

Technology Transfer & Commercialization Network

To meet the technological needs of American industry and boost U.S. international competitiveness, NASA operates a technology transfer network, composed of a National Technology Transfer Center (see page 124) and six Regional Technology Transfer Centers (RTTCs).

The RTTCs generally provide their clients a range of information, technical and commercialization services of similar nature, but each RTTC offers certain specialized services and each has close relationships with a particular NASA center or centers. They are geographically located to provide an equal distribution of services throughout the U.S. The regional deployment of the centers and their alignments with the Federal Laboratory Consortium allows the RTTCs to work closely with federal, state and local programs in serving the technology-related needs of business and industry.

The six RTTCs in the national network include:

• Serving the Northeast, the Center for Technology Commercialization (CTC), Inc. is a not-for-profit corporation that helps industry firms obtain and commercialize NASA, defense, industrial and university-developed technologies. CTC provides technology acquisition, market identification, partnering and commercialization services and, like the other RTTCs, conducts a TAP-IN technology utilization effort designed to help defense-oriented companies redirect their products and services into commercial markets. The CTC Technical Information Center offers technology research, marketing research, patent research, document and patent delivery services. The center also operates an "inner network" of eight Satellite Technology Transfer Centers located in the six New England States, New York and New Jersey.

• Far West Regional Technology Transfer Center, located at the University of Southern California, offers help to industry in such areas as new product identification and marketing, licensing opportunities, business development, funds sourcing, organizational networking, needs assessment, technology problem solving and research/engineering assistance. Far West RTTC has also initiated the new TAP-IN

Chris Coburn (right), executive director of the Great Lakes Industrial Technology Center (GLITeC), outlines a center program to a client at a GLITeC open house.

program to aid defense conversion measures. Under this program, Far West RTTC (and other RTTCs) can assist companies in finding new markets, developing new products, finding funding sources, writing proposals and business plans, and acquiring new technologies.

• Great Lakes Industrial Technology Center (GLITeC), the Midwest RTTC, works with industry in the six-state Great Lakes region (Illinois, Indiana, Michigan, Minnesota, Ohio and Wisconsin). Located in Cleveland, Ohio, within a mile of Lewis Research Center, GLITeC has special access to Lewis technology and staff and the two centers have initiated several technology transfer programs to promote industrial use of Lewis technology in commercial applications. GLITeC offers the full range of RTTC services, including technology assistance, commercialization, technology packaging, industrial problem solving and TAP-IN. To transfer NASA technology and commercialize Lewis technology throughout the 50 states, GLITeC maintains a network of state affiliates and partners which provide complementary services. GLITeC is managed by Battelle Memorial Institute, the world's largest nonprofit independent research organization.

• Headquartered at the University of Florida College of Engineering, the Southern Technology Applications Center (STAC) has a special responsibility as the focal point of a technology

Project manager Priscilla Diem (left) explains a joint GLITeC/ Lewis Research Center surface texturing program to a prospective client. Located a mile apart, GLITeC and Lewis have a close technology transfer relationship.

Ron Thornton (right) and Erik Sander (center) of the Southern Technology Applications Center learn about precision molded optics from their client, Dr. Jean-Luc Nogués of Geltech.

transfer partnership that involves three NASA field centers: Marshall Space Flight Center (Alabama), Kennedy Space Center (Florida) and Stennis Space Center (Mississippi). The partnership is known as the Southeast Technology Transfer Alliance; its mission is to provide direct access to NASA-generated technologies and indirect access to technologies available through the national technology transfer network. The Alliance enables private sector firms to use technologies, facilities and expertise for industrial problem solving, new product development and technology commercialization. Additionally, STAC performs the full range of technology transfer and commercialization services common to all the RTTCs. The STAC Information Services Center offers access to more than 1,000 databases worldwide and provides customized value-added searches. STAC works closely with the Federal Laboratory Consortium to enable private sector clients to access the extensive R&D library of some 700 federal laboratories.

• Mid-Continent Technology Transfer Center (MCTTC), headquartered at the Texas Engineering Extension Service (TEEX) of the Texas A&M University System, embraces a 14-state area served through a team of affiliates composed of private industry, university and federal/state agencies. The MCTTC has two primary customers: the federal laboratories and high tech industrial firms that need to acquire and commercialize new technology. By

exercising its unique position at TEEX, MCTTC extends its technology transfer resources to other programs, particularly the Texas Manufacturing Assistance Center and the Economic Development Administration at Texas A&M. MCTTC's range of services includes technical needs assessment, technology search, innovation process situation analysis, market assessment, applications engineering, TAP-IN, technology transfer agreement facilitation, company diagnostics and a variety of training courses.

• Located at the University of Pittsburgh, the Mid-Atlantic Technology Applications Center (MTAC) helps U.S. companies improve their competitiveness by assisting them in the location, assessment, acquisition and utilization of technologies and scientific/engineering expertise within the federal laboratory system. MTAC enables companies to stretch tight R&D budgets by helping them assemble a "virtual R&D" capability from external resources, such as the federal laboratories. The center works one-on-one to provide a variety of services supporting technology commercialization, including information retrieval, technical analyses and assessments market intelligence, product enhancement and applications development. MTAC has close associations with two NASA field centers: Langley Research Center and Goddard Space Flight Center.

(Continued)

Technology Transfer & Commercialization Network *(Continued)*

The hub of the national technology transfer network is the National Technology Transfer Center (NTTC). Located at Wheeling Jesuit College, Wheeling, West Virginia, NTTC serves as a clearinghouse for federal technology transfer, linking U.S. firms with federal agencies and laboratories, the RTTCs, and state/local agencies.

The center operates, in the interest of enhanced U.S. competitiveness, a "gateway service," toll-free telephone access to a full federal technology database and indexing system. By calling a 1-800 number, U.S. companies can access the federal laboratory system in search of technologies and research data that can assist them in developing their businesses.

The NTTC also provides training and educational services to government and industry to develop the skills essential to effective technol-

ogy transfer, and it conducts outreach and promotional activities to improve private sector awareness of technology transfer opportunities.

A new resource of the national network, introduced in 1996, is a special client/server database known as TechTracS, which is designed to monitor network-wide technology transfer activities. TechTracS is a means of inventorying and managing the great many technological projects that have commercialization potential; additionally, it provides a system for the administration and processing of inventions under the NASA Patent Program. The database links the 10 NASA field centers in a client/server structure that communicates across the Internet on a regular basis with the main database server at NASA Headquarters in Washington, D.C. TechTracS was developed by a team that included Research Triangle Institute, Research Triangle Park, North

The headquarters of the National Technology Transfer Center, the hub of the nationwide system, located at Wheeling (West Virginia) Jesuit College.

Walter M. Heiland heads the Technology Transfer Office at NASA's Center for AeroSpace Information.

Carolina and ACI US, Inc., Cupertino, California, developer of the 4th Dimension (4D) software that is the core of the system.

Support for all the elements of the National Technology Transfer Network is provided by the Technology Transfer Office at the Center for AeroSpace Information (CASI). This office executes a wide variety of tasks, among them maintenance of a document request list for and mailout of Technical Support Packages (TSPs), which provide details of new technologies available for more than 70 percent of the listings published in *NASA Tech Briefs* (see page 126). The mailout of TSPs involves a reproduction effort of more than 1.8 million pages annually. CASI is also responsible for responding to requests for information, an activity that entails processing of some 40,000 letters and other inquiries and mailout of about 300,000 documents a year. The office additionally

serves as a "central call-in" facility, channeling information and technical assistance seekers to the proper NASA technology transfer and commercialization organization or to other appropriate agencies.

In addition, the CASI Technology Transfer Office is responsible for research, analysis and other work associated with this annual *Spinoff* volume, for distribution of technology transfer publications, for retrieval of technical information and referral of highly detailed technical requests to appropriate offices, for developing reference and biographical data, and for public relations activities connected with media, industry and trade show interest in technology transfer matters and commercialization.

At the Center for AeroSpace Information, Jennifer Munro (standing), Maria Zimmerman (center) and Sharleen Angyelof research technology transfer cases for Spinoff; below, Lenora Parris (left) and Diane Odachowski prepare cutsheets for the NASA Tech Briefs publication.

Publications

An essential measure in promoting greater use of NASA technology is letting potential users know what technologies are available for transfer. This is accomplished primarily by the publication *NASA Tech Briefs*.

The National Aeronautics and Space Act requires that NASA contractors furnish written reports containing technical information about inventions, improvements and innovations developed in the course of work for NASA. These reports provide the input for *NASA Tech Briefs.* Issued monthly, the free publication is a current awareness medium and problem solving tool for more than 200,000 government and industry readers.

Each issue contains information on newly-developed products and processes, advances in basic and applied research, improvements in shop and laboratory techniques, new sources of technical data and computer programs, and other innovations originating at NASA field centers or at the facilities of NASA contractors.

An example of how *NASA Tech Briefs* inspires secondary application of NASA technology is the experience of Intelligent Vision Systems,

Inc. (InVision™), Houston, Texas, which ran into problems while developing the TDS-200 Traffic Detection System. Designed to monitor road traffic, the system consists of a series of pole-mounted sensors that identify shapes (vehicles, pedestrians), detect movement or the lack of it, count individual objects in their respective lanes, and calculate their speed; information of this type is important to highway control engineers. The problem that surfaced during development of the system was the sensor's inability to provide adequate image recognition in rain, fog or other bad weather.

InVision president Paul Mayeaux credited NASA technology with the breakthrough that solved the problem. He said: "After three years of limited R&D success and dwindling enthusiasm, our research group realized that we had to find another image-sensing approach. We had an economical computer, super signal processing hardware and software, but poor imagery."

In an issue of *NASA Tech Briefs,* an InVision researcher found an article describing a NASA technology developed for satellite imaging that

A technician of Intelligent Vision Systems, Inc. conducts a test of the company's TDS-200 Traffic Detection System. An article in NASA Tech Briefs enabled solution of a major problem with the TDS-200's sensory system.

utilizes multiple electromagnetic frequencies to improve image acquisition in all weather conditions. The InVision group requested and received from NASA a Technical Support Package (TSP), a collection of detailed technical data about the technology in question. The TSP, plus advice and consultation provided by Johnson Space Center, enabled solution of the problem after modification of the technology to meet InVision's special design requirements. The TDS-200 system is being produced commercially and it is in operational service at various U.S. locations.

Another example of *Tech Briefs* utility is supplied by Eliot Fenton, president of Integrated Component Systems, Inc. (ICS), Coconut Creek, Florida, which designs and manufactures synthesizers and oscillators used in wireless systems, modems, test equipment and related products. ICS products range in size from inch-square surface mount units to full size boards for frequency-hopping communications systems.

A regular reader of *Tech Briefs*, Fenton read an article therein that seemed to be the answer to an ICS problem. One of the company's custom-

ers wanted a synthesizer with wide phase modulation characteristics, yet low noise. The problem was that the two requirements are inherently incompatible in a single-loop design. An ICS engineer suggested an approach, but Fenton wanted confirmation.

The *Tech Briefs* article described research at Jet Propulsion Laboratory (JPL) resulting in a modified configuration for a phase-locked angle modulator that made it possible to design the filters in the modulating portion of the circuit independently of the filter in the phase-locked loop portion; applied to a phase-locked oscillator, it offered superior phase noise performance.

"The article gave my engineer valuable insight into how the process works," Fenton said, "and it substantiated our method as viable for wideband phase modulation." The technology was incorporated in the ICS series of phase-locked loop synthesizers.

[TM]InVision is a trademark of Intelligent Vision Systems, Inc.

A pair of representative ICS oscillators manufactured by Integrated Component Systems, Inc. The Tech Briefs publication gave the company a lead to an improvement that has been incorporated in the ICS line.

NASA's Technology Transfer & Commercialization Network

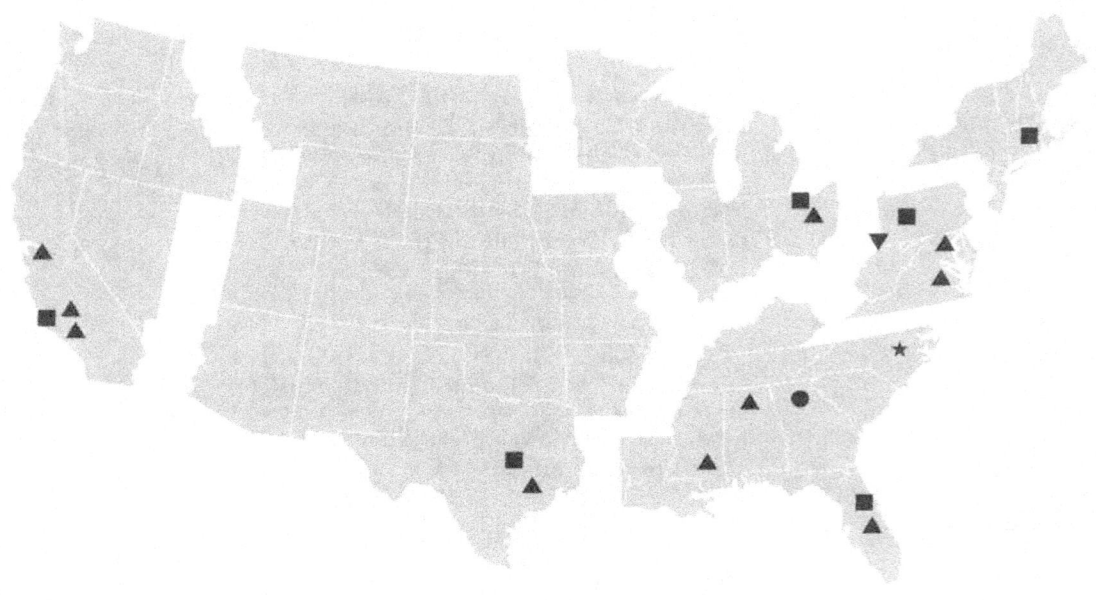

The NASA system of technology transfer and commercialization personnel and facilities extends from coast to coast. For specific information concerning the activities described below, contact the appropriate technology transfer personnel at the addresses listed, or address inquiries to the Manager, Technology Transfer Office, Center for AeroSpace Information, 800 Elkridge Landing Road, Linthicum Heights, Maryland 21090.

▲ **Field Center Technology Utilization Officers:** manage center participation in regional technology transfer activities.

▼ **National Technology Transfer Center:** national information, referral and commercialization service for NASA and other government laboratories.

■ **Regional Technology Transfer Centers:** information, technical and commercialization services.

● **The Computer Software Management and Information Center (COSMIC):** offers government-developed computer programs adaptable to secondary use.

★ **Application Team:** assists agencies and private institutions in applying aerospace technology to solution of public problems.

▲ FIELD CENTERS

Ames Research Center
National Aeronautics and
Space Administration
Moffett Field, California 94035
Director, Office of Commercial Technology:
Syed Z. Shariq, Ph.D
Phone: (415) 604-0753

Goddard Space Flight Center
National Aeronautics and
Space Administration
Greenbelt, Maryland 20771
Technology Transfer Officer:
George Alcorn, Ph.D.
Phone: (301) 286-5810

Lyndon B. Johnson Space Center
National Aeronautics and
Space Administration
Houston, Texas 77058
Director, Technology Transfer and
Commercialization Office:
Henry Davis
Phone: (713) 483-0474

John F. Kennedy Space Center
National Aeronautics and
Space Administration
Kennedy Space Center, Florida 32899
Technology Utilization Officer:
James A. Aliberti
Phone: (407) 867-3017

Langley Research Center
National Aeronautics and
Space Administration
Hampton, Virginia 23681-0001
Director, Technology Applications Group:
Joseph S. Heyman, Ph.D.
Phone: (804) 864-6005

Lewis Research Center
National Aeronautics and
Space Administration
21000 Brookpark Road
Cleveland, Ohio 44135
Chief, Commercial Technology Office:
Ann Heyward
Phone: (216) 433-3484

George C. Marshall Space Flight Center
National Aeronautics and
Space Administration
Marshall Space Flight Center, Alabama 35812
Technology Transfer Officer:
Harry G. Craft, Jr.
Phone: (205) 544-5418

Jet Propulsion Laboratory
4800 Oak Grove Drive
Pasadena, California 91109
Technology Transfer Office Manager:
Merle McKenzie
Phone: (818) 354-2577

NASA Management Office—JPL
4800 Oak Grove Drive
Pasadena, California 91109
Technology Commercialization Officer:
Arif Husain
Phone: (818) 354-4862

John C. Stennis Space Center
Mississippi 39529
Technology Transfer Officer:
Kirk Sharp
Phone: (601) 688-1914

Dryden Flight Research Facility
National Aeronautics and
Space Administration
Post Office Box 273
Edwards, California 93523-0273
Chief, Technology and
Commercialization Office:
Lee Duke
Phone: (805) 258-3802

■ REGIONAL TECHNOLOGY TRANSFER CENTERS
1-800-472-6785 You will be connected to the
RTTC in your geographical region.

Far-West
Technology Transfer Center
University of Southern California
3716 South Hope Street, Suite 200
Los Angeles, California 90007
Carolyn Suckow, Acting Director
Phone: (213) 743-2955
(800) 642-2872 (toll-free US)

Northeast

Center for Technology Commercialization
1400 Computer Drive
Westborough, Massachusetts 01581
William Gasko, Ph.D., director
Phone: (508) 870-0042

Mid-West

Great Lakes Industrial Technology Center
25000 Great Northern Corp. Ctr., Suite 260
Cleveland, Ohio 44070-5331
Christopher Coburn, director
Phone: (216) 734-0094

Southeast

Southern Technology Application Center
University of Florida
College of Engineering
Box 24
One Progress Boulevard
Alachua, Florida 32615-9987
J. Ronald Thornton, director
Phone: (904) 462-3913

Mid-Continent

Texas Engineering Extension Service
Texas A&M University System
301 Tarrow Street
College Station, Texas 77843
Gary Sera, director
Phone: (409) 845-8762

Mid-Atlantic

University of Pittsburgh
823 William Pitt Union
Pittsburgh, Pennsylvania 15260
Lani Hummel, director
Phone: (412) 648-7000
(800) 257-2725 (toll-free US)

● COMPUTER SOFTWARE MANAGEMENT AND INFORMATION CENTER

COSMIC
382 E. Broad Street
University of Georgia
Athens, Georgia 30602
Tim Peacock, director
Phone: (706) 542-3265

★ TECHNOLOGY APPLICATION TEAM

Research Triangle Institute
Post Office Box 12194
Research Triangle Park,
North Carolina 27709
Doris Rouse, Ph.D., director
Phone: (919) 541-6980

▼ NATIONAL TECHNOLOGY TRANSFER CENTER

Wheeling Jesuit College
Wheeling, West Virginia 26003
Ismail Akbay, executive director
Phone: (304) 243-2455
(800) 678-6882

NASA CENTER FOR AEROSPACE INFORMATION

Technology Transfer Office
800 Elkridge Landing Road
Linthicum Heights, Maryland 21090
Walter Heiland, manager
Phone: (301) 621-0241

Spinoff Team

Project Manager:
Walter Heiland

Senior Technical
Information
Specialist:
Jennifer Munro

Technical
Information
Associate:
Mindy Murdza

Graphic Services:
Steve Chambers
Creative Services

Photography:
Kevin Wilson

www.ingramcontent.com/pod-product-compliance
Lightning Source LLC
Chambersburg PA
CBHW081726170526
45167CB00009B/3717